Programming in Mathematica®

Programming in Mathematica®

Roman E. Maeder

Addison-Wesley Publishing Company, Inc.
The Advanced Book Program

Redwood City, California • Menlo Park, California
Reading, Massachusetts • New York • Amsterdam
Don Mills, Ontario • Sydney • Bonn • Madrid
Singapore • Tokyo • San Juan • Wokingham, United Kingdom

UNIX is a registered trademark of AT&T.
Mathematica is a registered trademark of Wolfram Research, Inc.
Macintosh is a trademark of Apple Computer, Inc.
POSTSCRIPT is a trademark of Adobe Systems Incorporated.
Sun-3 and Sun-4 are trademarks of Sun Microsystems, Inc.
SunView is a trademark of Sun Microsystems, Inc.
TeX is a trademark of the American Mathematical Society.
MIPS is a trademark of MIPS Computer Systems, Inc.
MS-DOS is a registered trademark of Microsoft Corp.

Publisher: *Allan M. Wylde*
Production Manager: *Jan Benes*
Promotions Manager: *Laura Likely*

Library of Congress Cataloging-in-Publication Data

Maeder, Roman.
    Programming in Mathematica / Roman Maeder.
    p.   cm.
    ISBN 0-201-51002-2
    1. Science–Data processing. 2. Programming (Electronic computers). I. Title.
    Q183.9.M34 1990   510'.28'553—dc19   90-295 CIP

This book was prepared by the author, using the TeX typesetting language on a Sun-3/160 computer.

ISBN 0-201-51002-2

BCDEFGHIJ-AL-943210

■ Preface

Mathematica was officially announced about a year ago. Since then it has found many uses in very diverse fields. While it is useful to do a few calculations interactively, its real strength lies in the programming language it offers. Writing programs, one extends *Mathematica* with specialized new functions in one's own field of interest. *Mathematica*'s programming language is unlike any you have encountered before. The manual explains all its features and gives some very basic examples of their use. For writing good programs, however, this in not enough and there is clearly a need for a book explaining all the many features in context and giving larger examples of their use. It is my hope that this book fills this gap.

I have been in contact with many early users of *Mathematica* and I have also given students at the University of Illinois an opportunity to learn about it. I have seen many programs written in *Mathematica*, good ones and bad ones. The bad ones invariably solve a particular problem in an unnecessarily complicated way, unaware of the elegant constructs available in *Mathematica*. Let me point out these constructs and show how they can be used to write concise, elegant, and efficient programs.

The best way to teach about these *Mathematica*-specific programming methods, is to look at examples of complete programs that solve some non-trivial problem. Even if the example chosen does not lie in the particular field of application that you are interested in, you will be able to use similar ideas for your own programs and, therefore, hopefully find it useful. Many of the examples presented here deal with graphics. Graphics applications are especially suited for learning programming, since it is easy to see whether your code is correct simply by looking at the picture it produces. Advanced graphics applications are also sufficiently challenging to require advanced programming methods. Other examples come from symbolic computation and numerical mathematics. Along the way, we also develop many short pieces of code that can be used as parts of larger programs or that help to customize *Mathematica* to your particular needs.

Since there exists an excellent manual for *Mathematica* itself, this book does not explain everything from scratch, but assumes some familiarity with *Mathematica*. I assume also that you have access to a computer on which to try out the examples. You might want to obtain the examples from this book in machine-readable form, since it is rather pointless to type them in again. I already did this for you.

I am grateful to many people that have contributed to this book. My thanks go first to the other developers of *Mathematica*. The language was shaped through countless discussions and many heated arguments. Trying to explain to each other *why* we think a certain feature should be done in a certain way has deepened our understanding of the matters involved and has given the language its overall consistency, despite the fact that it contains hundreds of commands and unifies many diverse programming paradigms.

Helpful ideas for this book came from Theodore Gray, Dan Grayson, Ferrell Wheeler, and Stephen Wolfram. Some of the examples I used were inspired by Henry Cejtin, John Gray, Lee Rubel, William Thurston, and Dave Withoff. I would also like to thank the students of Math 351 (Spring term 1989) at the University of Illinois at Urbana-Champaign, who were the first to be introduced to the material contained in this book.

Help with the typographical side of producing this book came from John Bonadies, Daniel Lee, Cameron Smith, and Gregg Snyder, as well as from many other people at Wolfram Research, Inc. and the staff of Addison-Wesley, in particular Allan Wylde and Jan Benes.

R. E. M.

Urbana, Illinois
June, 1989

A computer, to print out a fact,
Will divide, multiply, and subtract.
But this output can be
No more than debris,
If the input was short of exact.
– Gigo

■ Contents

■ 9 Input and Output

■ 10 Notebooks

■ 11 Producing the Cover Picture

■ Appendix A Exercises

■ Appendix B Program Listings

■ Index

■ About This Book

In this book I want to tell you how you can use *Mathematica* for more than just typing in single commands. If you are using *Mathematica* for your work, teaching or solving homework assignments, sooner or later you will encounter more advanced problems requiring many commands to solve in *Mathematica*, or you are faced with doing the same calculation steps over and over again with different input. This is the point at which you want to start writing *programs* in *Mathematica*'s programming language.

Mathematica includes a rich and powerful programming language. Unlike the usual languages like BASIC, C etc. it is not restricted to having a small number of data types but allows you to perform all symbolic computations of *Mathematica*. The *Mathematica* manual can only hint at the possibilities. It explains all the features but does not show you which ones to use for a particular problem or how to fit things together into larger programs.

Through design and tradition each programming language has developed a certain preferred style of good programming. It is possible to solve the same problem in many different ways, but there is usually some idea about what is a good or a bad program. *Mathematica* is new and there are, as yet, few good programs that have established such a style. In this book, I want to present examples of what the designers of *Mathematica* think is good programming style and show you why this is so. Even if you write your programs strictly for personal use, you will benefit from following a good style. For developing programs for others to use, adhering to this style is indispensible.

■ How the Examples are Developed

The programming examples in this book serve two purposes. One, they help explain concepts and show how things fit together to complete programs. Two, they are designed to be more than mere toy programs and should prove useful in their own right.

In developing an example we always use the same method. We start out with a few commands or definitions that could be entered directly into *Mathematica*. We then extract the parts of the computation that are the same regardless of the input and define some functions or procedures that automate these steps. Then we apply standard techniques to these functions to make them into a *package*, adding documentation and certain programming constructs that make such a package easier to use. The goal is to write a program that would be useful not only to its author who knows how it works, but also to other people. Finally we might add some more functions to the package or look at alternatives to what we did so far.

In Chapter 1 these steps are described in full detail. Later on when we concentrate on other aspects of *Mathematica*, we assume that you are familiar with these basic concepts and will not mention all the steps in detail.

A package is identified by a name (the context name, as we will see), for example ComplexMap. The files in which we store successive versions of this package will be called **ComplexMap1.m**, **ComplexMap2.m**, and so on. The final version will then simply be called **ComplexMap.m**. Much of the text in successive versions of the same package is the same and we do not generally reproduce it in full. The final package is reproduced in full in Appendix B.

■ Contents of the Chapters

Chapter 1 develops a package from scratch. It follows the steps outlined in the previous subsection. Along the way we learn how to set up package contexts, define defaults for parameters of functions, and also look at graphics. The example chosen comes from mathematics: functions of one complex variable. You can follow the material even if you are not familiar with the mathematical concepts.

In Chapter 2 we look at the theory of writing good packages in *Mathematica*. The concepts introduced here will be used throughout the rest of the book. Understanding why things should be done in a certain way might be interesting on its own, but you can also learn to apply the concepts in a cookbook style way. A skeletal package is provided, which you can use as a template for writing your own packages.

Chapter 3 looks at issues of pattern matching and defining options for functions that you define. Default values for parameters and options can make commands much easier to use. *Mathematica* itself makes heavy use of this.

Chapter 4 looks at different programming styles possible in *Mathematica*. It is here that the preferred functional style is explained. Understanding this chapter will allow you to write better programs.

Chapter 5 looks at some aspects of how *Mathematica* evaluates expressions. If you want to write sophisticated rules and definitions you need to know about these things. Otherwise you can skip this material and return to it later as needed.

In Chapter 6 we introduce "mathematical programming," a way of expressing mathematical relations and formulae that is unique to *Mathematica*. If your application of *Mathematica* requires simplification and transformation of symbolic expressions, you should read this chapter.

Numerical computations are the topic of Chapter 7. It explains how numerical computations are done and what kind of numbers *Mathematica* supports. You should consult it as needed.

Mathematica has certain built-in rules and procedures for performing certain transformations automatically. Sometimes it is necessary to change these built-in rules or add others. Chapter 8 explains how this can be done. You should have read Chapter 5 before reading the material in here.

Chapter 9 treats input and output. *Mathematica* interacts with the rest of your computer system in many ways and this chapter deals with this interface.

Chapter 10 looks at *notebooks*, available on some frontends of *Mathematica*. It looks at the relationship between notebooks and packages. If you use notebooks or if you want to write programs that run on machines with or without the notebook interface, you should read this chapter.

The last chapter explains how the picture on the book cover was generated. Along the way we develop some useful graphics utilities and an improved version of **ParametricPlot3D.m**.

Appendix A contains some exercises and solutions. Appendix B contains the program text of all examples used in this book that are not reproduced in full in one of the chapters.

Some sections are marked as "Applications." They introduce a programming example that makes use of the concepts covered in the preceding sections. They are independent of the rest of the text. Some sections are marked "Advanced Topic." They have a higher level of difficulty than the rest of the book or require some higher mathematics. They, too, are optional.

■ References to the *Mathematica* Book

This book is no replacement for the *Mathematica* manual "*Mathematica*: A System for Doing Mathematics by Computer", by Stephen Wolfram (Addison-Wesley, 1988), referred to as the "*Mathematica* book." One copy of it comes with every copy of *Mathematica*. I do not expect that you have read everything in the *Mathematica* book but you should have some basic experience with *Mathematica* before reading this book. Single commands are usually used without explaining them in detail. You can use the index in the *Mathematica* book to look up a description of a command that you did not know about. We also give references to places in the *Mathematica* book where you can find explanations of concepts that are particularly relevant to a topic in this book. You should always turn to the *Mathematica* book for explanations of features that are assumed known here, but that you have not used yet. The place to look for an explanation about all variants, defaults, or options for a particular command is the *Reference Guide* in the back of the *Mathematica* book.

■ Trying Out the Examples

All examples were tested with version 1.2 of *Mathematica*. The "live" calculation sequences in this book were computed on a Sun-3 computer running *Mathematica* 1.2 and SunOS 4.0.1. Some examples will not work with earlier versions of *Mathematica*. No attempt has been made to identify them. If you want to reproduce these calculations and add your own (which I suggest you do while reading this book) you should use version 1.2 or later. Those examples which interact with the operating system of the computer on which *Mathematica* is running are, of course, machine dependent and will look completely different on computers that do not run UNIX.

Some of the packages in this book depend on other packages. In order to read them into *Mathematica*, all these imported packages must be available too. Some of these are improved versions of standard packages and have the same names. Make sure that the correct packages are read in.

In the example *Mathematica* sessions in the later chapters, we will generally no longer show the command to read in the package that is the topic of the example. This command of the form <<*Package*.m is assumed at the beginning of every session that uses functions from the package.

All examples are available in machine-readable form in a variety of formats, see Appendix B. Please note that they are protected by copyright and may not be distributed further except as permitted by the copyright notices they contain.

■ Which Version of *Mathematica* is Described?

All explanations about how *Mathematica* works are based on Version 1.2. Some features are not available in earlier versions. Note that the current edition of the *Mathematica* book describes Version 1.1. You should consult the upgrade documentation that came with your copy of Version 1.2 of *Mathematica* for information about any differences and new features.

This book is about programming, and is concerned only with the kernel of *Mathematica*, not with any of the various frontends available. The material is therefore useful for all versions of *Mathematica*. Only in Chapter 10 will we talk about the interaction between the programs developed here and the notebooks available in some versions.

■ Notation and Terminology

Mathematica input and output is typeset in typewriter-like style: Expand[(x+y)^9]. Parts of such input that are not literal, but denote (meta-)variables, are typeset in italics: f[*var_*] := *body*. *Functions* or *commands* are referred to by their name followed by an empty argument list, for example Expand[]. Pieces of *Mathematica* programs are delimited with horizontal lines and usually have captions beneath them.

```
a[1] = a[2] = 1
a[n_Integer?Positive] := a[n] = a[a[n-1]] + a[n-1-a[n-1]]
```

John H. Conway's $10,000.- sequence

Names of files containing *Mathematica* programs are typeset in this boldface style: **ParametricPlot3D.m**. Genuine dialogue with *Mathematica* is set in two columns. The left column contains explanations and the right column contains the *Mathematica* input and output, including graphics. You should be familiar with these conventions from the *Mathematica* book.

In the programming examples, I have tried to follow a uniform style for the indentation of lines in a definition. Since *Mathematica* allows you to write deeply nested expressions, lines are often rather long and have to be broken up so as to fit on the printed page.

In most programming languages you can define *procedures*, *subroutines*, or *functions*. In *Mathematica*, all of these are just another way of looking at *definitions*, commands of the form $f[x_-] := body$. Often we also use the term *global rule*. These terms will be used interchangeably, depending on which point of view we want to stress in a particular situation. Rules proper are things of the form *lhs -> rhs* or *lhs :> rhs*. A *substitution* is an expression of the form *expr /. rule*.

Chapter 1

From Interactive Commands to a Package

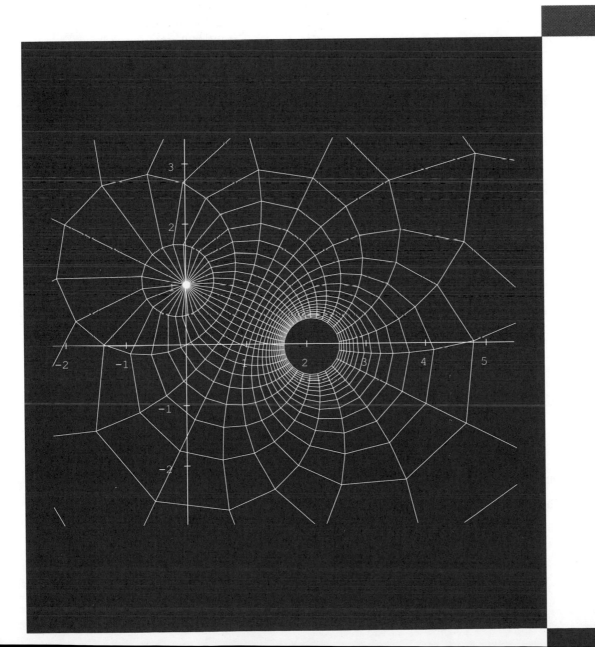

In this chapter we will develop a sample package from scratch. We start out with commands entered interactively into *Mathematica*. Then we collect them into a definition in which the parts we would like to modify are defined as parameters to a replacement rule. This rule will look like a traditional procedure definition in any of the common higher level programming languages.

If you save this definition in a file to read it into *Mathematica* by a single command you have written your first "package." We will then add commands to set up a separate context for the functions defined in the package. This will isolate any local variables and auxiliary functions used in the implementation of the package. It will make only those objects visible that are to be exported from the package. These are the functions and variables that you can use after having read the package into your current *Mathematica* session.

Next, a second function is added to the package. Some of the code common to both functions will be put into a separate auxiliary function. This saves space and makes the program easier to maintain. If a change is required, it has to be done in only one place and will be consistently used everywhere.

Then it will be time to add some useful extra features to the basic algorithms. We will define default values for frequently used parameters. Another area of concern is the graceful handling of bad user input to one of our functions. We will deal with arithmetic exceptions like division by zero.

The example chosen for this chapter comes from mathematics. We will draw graphs of complex-valued functions. However you do not need to know much about the mathematical properties of complex numbers to understand this example. The emphasis is on the programming part.

Sections 1.1 through 1.3 will get you from the interactive session to a fully developed package that is useful in its own right. The following sections will introduce refinements that should be done for any package that you want to use often or make available to others.

About the illustration overleaf:

A picture of a Möbius transform generated with the command

```
PolarMap[(2#-I)/(#-1)&, {0.001, 5.001, 0.25}, {0, 2Pi, Pi/15}].
```

This command (it is not built-in) will be developed in this chapter.

■ 1.1 Collecting Commands in a Package

This chapter shows the stages in building up a package in *Mathematica*. To show how things really work, we will use a real example, complete with all the necessary details. The package we choose is one for plotting functions of complex variables. Built into *Mathematica* is the Plot[] function (see section 1.8 of the *Mathematica* book), which can plot functions of one real variable.

Most of the functions (logarithms, exponential function, trigonometric, and inverse trigonometric functions) encountered in high school and college mathematics can be evaluated for complex-valued arguments as well. *Mathematica* does not have a built-in command to plot functions of a complex variable. Therefore let us add one!

A complex number consists of two quantities, its *real* and *imaginary* part. These can be thought of as the coordinates of a point in the plane. The complex number describing the point with coordinates (a, b) is written $a + ib$. i is the imaginary unit; it stands for the square root of -1. In *Mathematica* you write I instead of i since all built-in names start with capital letters.

■ 1.1.1 Plotting the Coordinate Lines in the Complex Plane

Since plotting a function of a complex variable would need four dimensions (two for the complex variable and two for the complex function value) we need a different way of visualizing such a function. One way to see how the function transforms the complex plane is to look at the image of the coordinate lines, the horizontal lines of constant imaginary part, and the vertical lines of constant real part. Each of these lines will be mapped into some curve in the complex plane and we can plot these curves in two dimensions.

In order to draw the images of the coordinate lines under the given function, the sine function for example, we compute the function values at the intersection points of these lines and then generate Line[] commands to connect the points with horizontal and vertical lines.

First, let us draw the coordinate lines themselves. We generate a two-dimensional table of the complex numbers corresponding to the intersection points of the grid and then replace each one with its coordinates. The coordinates of a complex number z are its real and imaginary parts. These are obtained by the *Mathematica* functions Re[z] and Im[z].

This generates a 15 by 11 table of complex numbers. As the output is rather long (and boring), we suppress it by placing a semicolon at the end of the command.

```
In[1]:= points = Table[ N[x + I y],
          {x, -Pi/2, Pi/2, Pi/14}, {y, -1, 1, 2/10} ];
```

But we can verify the correctness of the result by looking at a short outline of it. (The form of the output of Short[] depends a great deal on the line width and may be different on your computer.)

```
In[2]:= Short[points, 4]
Out[2]//Short=
 {{-1.5708 - 1. I, -1.5708 - 0.8 I, -1.5708 - 0.6 I,
     -1.5708 - 0.4 I, -1.5708 - 0.2 I, -1.5708,
     -1.5708 + 0.2 I, <<1>>, -1.5708 + 0.6 I,
     -1.5708 + 0.8 I, -1.5708 + 1. I}, <<14>>}
```

We replace each complex number by a list of its real and imaginary parts. The numbers are at level 2 in the two-dimensional table points. See section 5.2 for an explanation of how the "pure function" {Re[#], Im[#]}& works.

```
In[3]:= coords = Map[ {Re[#], Im[#]}&, points, {2} ];
```

Now we have a matrix of coordinate pairs. The x-coordinate of each point is the real part of the complex number and the y-coordinate is its imaginary part.

```
In[4]:= Short[coords, 3]
Out[4]//Short=
 {{{-1.5708, -1.}, {-1.5708, -0.8}, {-1.5708, -0.6},
     {-1.5708, -0.4}, {-1.5708, -0.2}, <<4>>,
     {-1.5708, 0.8}, {-1.5708, 1.}}, <<14>>}
```

The vertical lines are just the rows of the table.

```
In[5]:= vlines = Map[ Line, coords ];
```

The horizontal lines are the columns of the table or the rows of the transposed table.

```
In[6]:= hlines = Map[ Line, Transpose[coords] ];
```

Finally we join the two sets of lines and make a graphics object out of them.

```
In[7]:= Show[ Graphics[ Join[hlines, vlines] ],
           AspectRatio->Automatic ]
```

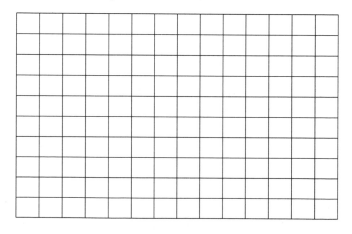

■ 1.1.2 A Picture of the Sine Function

If we replace the numbers (x + I y) by Sin[x + I y] in the previous computation we immediately get the net of the coordinate lines under the action of the sine function.

This generates 15 by 11 values of the sine function in the complex plane.

```
In[1]:= points = Table[ N[Sin[x + I y]],
            {x, -Pi/2, Pi/2, Pi/14}, {y, -1, 1, 2/10} ];
```

```
In[2]:= coords = Map[ {Re[#], Im[#]}&, points, {2} ];
```

This shows a slightly different way of computing the two sets of lines.

```
In[3]:= lines = Map[ Line,
            Join[coords, Transpose[coords]] ];
```

This renders the lines and also adds axes to the picture.

```
In[4]:= Show[ Graphics[lines], AspectRatio->Automatic,
            Axes->Automatic ]
```

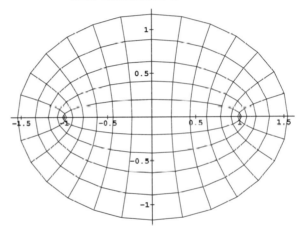

■ 1.1.3 Putting the Commands into a Procedure

If we want to generate graphs of the coordinate lines under different functions it makes
sense to collect the necessary commands in a procedure so we don't have to enter them
every time. The variable parts of the computation, the name of the function to be plotted,
and the ranges of the coordinates, are defined as parameters to this procedure.

```
CartesianMap[ func_, {x0_, x1_, dx_}, {y0_, y1_, dy_} ] :=
    Block[ {x, y, coords, lines},
        coords = Table[ N[func[x + I y]], {x, x0, x1, dx}, {y, y0, y1, dy} ];
        coords = Map[ {Re[#], Im[#]}&, coords, {2} ];
        lines = Map[ Line, Join[coords, Transpose[coords]] ];
        Show[ Graphics[lines], AspectRatio->Automatic, Axes->Automatic ]
    ]
```

CartesianMap1.m: The first version of `CartesianMap[]`

A few explanations:

- All the variables local to `CartesianMap[]` are declared in the `Block[]` statement
  to isolate them from any values they might have globally. It is important that the
  two variables `x` and `y` do not have values inside `CartesianMap[]` as they are
  used as variables in an iterator.

- The first argument of `CartesianMap[]` is the name of the function to be plotted. It
  is later used as the head of an expression in `func[x + I y]`. If we want to specify
  a function that is not built in we can either write a definition for it beforehand or
  use a pure function as the argument.

- Note the form of the patterns used for matching the two ranges. Each of them
  matches a list of exactly three elements. Since we do not need the list as a whole
  we give names to the individual elements only.

Here is an example of using this definition. We draw a map of the cosine function.

This reads in our file.

```
In[1]:= << CartesianMap1.m
```

This generates 20 by 17 lines. The lines do not close up because the range in the real direction is a bit smaller than π. The picture looks similar to the one of the sine function. Sine and cosine are very closely related in the complex plane.

```
In[2]:= CartesianMap[ Cos, {0.2, Pi-0.2, (Pi-0.4)/19},
                            {-2, 2, 4/16} ]
```

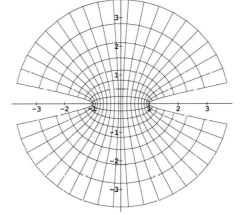

■ 1.2 The Basic Ingredients of a Package

In the previous section we wrote a file containing our first version of the function `CartesianMap[]`. While this file is already useful, it is not yet in a state to be published or made available to other users of *Mathematica*.

The goal in writing a *package* is to make the functions defined therein behave as much like built in *Mathematica* functions as possible. They should have documentation, accessible by typing `?CartesianMap`, and their behavior should not depend on the previous calculations that you have done in your *Mathematica* session before reading in the package. Several things can go wrong if you read a set of definitions into your current session:

- You could have defined values for variables that are used inside the definitions.
- You could pass variables as arguments that are also used locally inside the function.
- A function with the same name could already have been defined somewhere else.
- Auxiliary functions or private variables that are used inside the package would be accessible to the user. Users could rely on details of the implementation or make changes to it.

The following "package" `BadExample.m` illustrates one of these problems. The more subtle ones become only apparent in the context of a longer session with *Mathematica* and are then normally very hard to find.

```
(* this function returns the first n powers of x *)
PowerSum[x_, n_] :=
    Block[{i},
        Sum[ x^i, {i, 1, n} ]
    ]
```

BadExample.m: How not to write a package

We read in the file.

```
In[1]:= << BadExample.m
```

No problem so far. The output is as expected.

```
In[2]:= PowerSum[x, 5]

Out[2]= x + x^2 + x^3 + x^4 + x^5
```

The variable `i` is captured by the variable in the range of the summation. Instead of the expected `i + i^2 + i^3 + i^4 + i^5` we get a number.

```
In[3]:= PowerSum[i, 5]

Out[3]= 3413
```

■ 1.2.1 Putting Things in their Proper Context

The mechanism that *Mathematica* provides for keeping the variables used in a package different from those used in the main *Mathematica* session is called *contexts*. As each symbol is read from the terminal or from a file *Mathematica* checks to see whether this symbol has already been used before. If it has been encountered before, the new instance is made to refer to that previously read symbol. Otherwise you could not refer to the value of a variable you had just defined. If the symbol has not been encountered before, a new entry in the symbol table is created.

Each symbol belongs to a certain context. Within one context the names of symbols are unique, but the same name can occur in two different contexts. By default all new symbols that you define are put in the context Global` (note that context names always end with a `). The local variable i used in the definition of the function PowerSum above is therefore in the same context as the variable i used as argument to that function in input line 3.

If we tell *Mathematica* to create new symbols in a different context, we can avoid the problem.

```
PowerSum::usage = "PowerSum[x, n] returns the first n powers of x."

Begin["Private`"]

PowerSum[x_, n_] :=
    Block[{i},
        Sum[ x^i, {i, 1, n} ]
    ]

End[]
```

BetterExample.m: How to write a better package

The local variable i is now created in the context Private` which is not searched when you type in a variable name later on. The usage message defined for the symbol PowerSum is there not just to provide documentation for the function (which would be reason enough), but to make sure that the symbol PowerSum is defined in the current (global) context. If it had not been defined before entering the context Private` it, too, would not be found later on.

PowerSum[] should now do the expected thing even if the variable we give as parameter happens to be i. If you try this example for yourself, be sure to start a fresh *Mathematica* at this point, otherwise the previous definition would get in the way.

The value returned is the value of the command End[], which returns the name of the previous context.

```
In[1]:= << BetterExample.m
Out[1]= Private`
```

The variable i is not captured by the variable in the range of the summation.

```
In[2]:= PowerSum[i, 5]
Out[2]= i + i  + i  + i  + i
             2    3    4    5
```

The context of PowerSum is the default global context.

```
In[3]:= Context[PowerSum]
Out[3]= Global`
```

i is also in this context. The variable i inside PowerSum[] is inaccessible.

```
In[4]:= Context[i]
Out[4]= Global`
```

■ 1.2.2 A Package Context for CartesianMap

In addition to hiding local variables and functions, we also want to put all the functions that the package provides into a separate context. This context, however, must be visible so we can use the functions later on. This is achieved by the pair of commands BeginPackage[] and EndPackage[]. With these additions we present our second version of CartesianMap.m.

```
BeginPackage["CartesianMap`"]

CartesianMap::usage = "CartesianMap[f, {x0, x1, dx}, {y0, y1, dy}] plots
    the image of the cartesian coordinate lines under the function f."

Begin["`Private`"]

CartesianMap[ func_, {x0_, x1_, dx_}, {y0_, y1_, dy_} ] :=
    Block[ {x, y, coords, lines},
        coords = Table[ N[func[x + I y]], {x, x0, x1, dx}, {y, y0, y1, dy} ];
        coords = Map[ {Re[#], Im[#]}&, coords, {2} ];
        lines = Map[ Line, Join[coords, Transpose[coords]] ];
        Show[ Graphics[lines], AspectRatio->Automatic, Axes->Automatic ]
    ]

End[]
EndPackage[]
```

CartesianMap2.m: The second version of CartesianMap[]

Note the initial ` in the context name inside the command Begin["`Private`"]. This establishes `Private` as a subcontext of the context CartesianMap` (so its full name is CartesianMap`Private`).

We do not get an output line from reading in the file since `EndPackage[]` does not return a value.

```
In[1]:= << CartesianMap2.m
```

The function `CartesianMap[]` is in its own context.

```
In[2]:= Context[CartesianMap]
Out[2]= CartesianMap`
```

This context however is accessible because it has been put on the context search path.

```
In[3]:= $ContextPath
Out[3]= {CartesianMap`, Global`, System`}
```

The function works just as before.

```
In[4]:= CartesianMap[Exp, {-1, 1, 0.2}, {-2, 2, 0.2}]
```

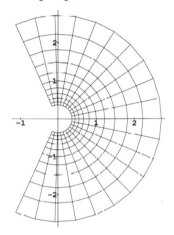

While developing and testing a new package, it is often a good idea to leave out the `BeginPackage[]` and `EndPackage[]` calls in the early stages when the code could still contain syntax errors. If a syntax error occurs in the middle of the package, *Mathematica* stops reading further commands from the package and the contexts will end up wrong. As soon as the syntax errors are removed, the context manipulating commands can be added.

■ 1.3 Adding Another Function to the Package

Writing a package for a single function is probably not worth the extra effort. We will now add a second function `PolarMap[]`. Instead of looking at the transformation of the Cartesian coordinate lines, we can also look at the transformation of the polar coordinate lines. In polar coordinates, each point in the complex plane is described by its distance from the origin (the radius or absolute value of the complex number) and the angle of its radius line measured from the positive x-axis (the argument of the complex number). If we know the radius and argument (r, ϕ) of a complex number z, we can easily compute its real and imaginary parts as $r \cos \phi$ and $r \sin \phi$ respectively. These two formulae can be expressed more concisely as $z = re^{i\phi}$.

As in section 1.1 we start out with a picture of the coordinate lines themselves.

The radius starts at 0. The angle ϕ goes once around the circle.

```
In[1]:= points = Table[ N[r Exp[I phi]], {r, 0, 1, 0.1},
                        {phi, 0, 2Pi, 2Pi/24} ];
```

The rest of the calculations is exactly the same as for the Cartesian coordinate lines.

```
In[2]:= coords = Map[ {Re[#], Im[#]}&, points, {2} ];
```

```
In[3]:= vlines = Map[ Line, coords ];
```

```
In[4]:= hlines = Map[ Line, Transpose[coords] ];
```

The lines of constant radius are circles around the origin. The lines of constant angle are straight lines radiating from the center.

```
In[5]:= Show[ Graphics[ Join[hlines, vlines] ],
              AspectRatio->Automatic ]
```

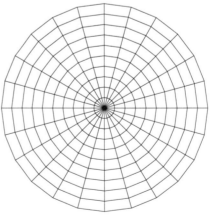

Since all the calculations are the same except that the complex numbers are generated as r Exp[I phi] instead of (x + I y), we can use the same code that we used for the function CartesianMap[]. We put this common code into a separate auxiliary function that is then used inside CartesianMap[] and PolarMap[]. This function should take a matrix of complex numbers as input and return a Graphics[] object that will plot the lines corresponding to the rows and columns of the matrix.

We put both functions into the same package, changing its name to ComplexMap. Here it is:

```
BeginPackage["ComplexMap`"]

CartesianMap::usage = "CartesianMap[f, {x0, x1, dx}, {y0, y1, dy}] plots
    the image of the cartesian coordinate lines under the function f."

PolarMap::usage = "PolarMap[f, {r0, r1, dr}, {phi0, phi1, dphi}] plots
    the image of the polar coordinate lines under the function f."

Begin["`Private`"]

MakeLines[points_] :=
    Block[ {coords, lines},
        coords = Map[ {Re[#], Im[#]}&, points, {2} ];
        lines = Map[ Line, Join[coords, Transpose[coords]] ];
        Graphics[lines]
    ]

CartesianMap[ func_, {x0_, x1_, dx_}, {y0_, y1_, dy_} ] :=
    Block[ {x, y, coords},
        coords = Table[N[func[x + I y]], {x, x0, x1, dx}, {y, y0, y1, dy}];
        Show[ MakeLines[coords], AspectRatio->Automatic, Axes->Automatic ]
    ]

PolarMap[ func_, {r0_, r1_, dr_}, {phi0_, phi1_, dphi_} ] :=
    Block[ {r, phi, coords},
        coords = Table[N[func[r Exp[I phi]]], {r, r0, r1, dr}, {phi, phi0, phi1, dphi}];
        Show[ MakeLines[coords], AspectRatio->Automatic, Axes->Automatic ]
    ]

End[(* "`Private`" *)]
EndPackage[]
```

ComplexMap1.m: Polar and Cartesian maps in the same package

The function MakeLines[] is private to the package. It cannot be accessed from outside, as it is defined in the private context of the package. An advantage of this is that the author of the package is free to change the implementation details of it without changing the way the package behaves.

In[1]:= << ComplexMap1.m

This is the image of the polar coordinate lines under the logarithm, which is the inverse of the exponential function that we have seen in section 1.2.2 on page 11.

In[2]:= PolarMap[Log, {0.1, 10, 0.5}, {-3, 3, 0.15}]

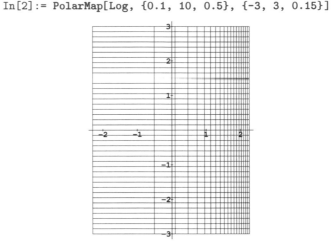

Here is a picture of the inversion at the unit circle. This function is not built-in, so we use a pure function instead. The formula for the inversion is $f(z) = 1/\bar{z}$ which is written as 1/Conjugate[#]& using the symbol # to denote the argument position of the pure function.

The images of the circles around the origin are again circles and the images of the radius lines are again radius lines. The spacing of the lines is no longer even, however (compare this with the image of the polar coordinate lines themselves earlier in this section). Note that we do not see all of the intersection points since some of them are rather far away from the origin and *Mathematica* automatically cuts off such points. The image of the origin is at infinity. Therefore we chose not to begin the radius at 0 but at the small positive value 0.1.

In[3]:= PolarMap[1/Conjugate[#]&, {0.1, 5.1, 0.5},
                                    {-Pi, Pi, 2Pi/24}]

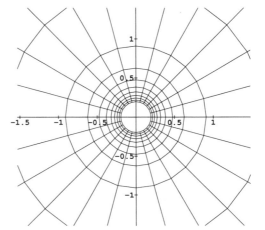

■ 1.4 Adding Defaults and Options

In both functions, `CartesianMap[]` and `PolarMap[]`, the ranges for the two variables are given as lists of three elements {*start, final, increment*}. It would be very convenient if the increment had a default value as it does in other iterator constructs. You do not want to use a fixed numerical value as default, but rather would like to specify the number of lines to draw and compute the increment accordingly. If you want to draw n lines between the values *start* and *final*, you compute the increment by dividing the difference of *start* and *final* by the number of lines to draw minus one (as an extra line is drawn at the end). So the formula is $(final - start)/(n - 1)$. What should the default value for n be? A good choice is the value of the option `PlotPoints` of the command `Plot3D[]` since this too, gives the number of lines to draw.

Default values that need to be computed cannot be given inside the pattern in the function definition, but rather we have to give a second rule that matches when a parameter is left out. When we leave out the two increments `dx_` and `dy_` in the parameter list for the function `CartesianMap[]`, we have a pattern that matches only when the two ranges given have two elements instead of three. In the body of this new rule we compute the necessary increment and then call `CartesianMap[]` again, this time with the increment. The other rule then matches and draws our picture.

```
CartesianMap[ func_, {x0_, x1_}, {y0_, y1_} ] :=
    Block[{dx, dy, plotpoints},
        plotpoints = PlotPoints /. Options[Plot3D];
        dx = (x1 - x0)/(plotpoints-1);
        dy = (y1 - y0)/(plotpoints-1);
        CartesianMap[ func, {x0, x1, dx}, {y0, y1, dy} ]
    ]
```

A rule for default increments

Recall that the value of `Options[Plot3D]` is a list of replacement rules, among them `PlotPoints -> 15` (there could be another number instead of 15 in your *Mathematica* version). Applying this list to the expression `PlotPoints` extracts the current value of this option, which is then saved in the variable `plotpoints`.

A similar rule is used for `ComplexMap[]`. As another convenience we give `r0` the default value 0. Most of the time we want to start the radius lines at the origin. Since this default is a simple number, we can put it right into the pattern as `r0_:0`. The updated package is in **ComplexMap2.m**. Note also the updated usage messages.

```
BeginPackage["ComplexMap`"]

CartesianMap::usage = "CartesianMap[f, {x0, x1, (dx)}, {y0, y1, (dy)}] plots
    the image of the cartesian coordinate lines under the function f.
    The default values of dx and dy are chosen so that the number of lines
    is equal to the value of the option PlotPoints of Plot3D[]"

PolarMap::usage = "PolarMap[f, {r0:0, r1, (dr)}, {phi0, phi1, (dphi)}] plots
    the image of the polar coordinate lines under the function f.
    The default values of dr and dphi are chosen so that the number of lines
    is equal to the value of the option PlotPoints of Plot3D[]"

Begin["`Private`"]

MakeLines[points_] :=
    Block[{coords, lines},
        coords = Map[ {Re[#], Im[#]}&, points, {2} ];
        lines = Map[ Line, Join[coords, Transpose[coords]] ];
        Graphics[lines]
    ]

CartesianMap[ func_, {x0_, x1_, dx_}, {y0_, y1_, dy_} ] :=
    Block[ {x, y, coords},
        coords = Table[N[func[x + I y]], {x, x0, x1, dx}, {y, y0, y1, dy}];
        Show[ MakeLines[coords], AspectRatio->Automatic, Axes->Automatic ]
    ]

CartesianMap[ func_, {x0_, x1_}, {y0_, y1_} ] :=
    Block[{dx, dy, plotpoints},
        plotpoints = PlotPoints /. Options[Plot3D];
        dx = (x1 - x0)/(plotpoints-1); dy = (y1 - y0)/(plotpoints-1);
        CartesianMap[ func, {x0, x1, dx}, {y0, y1, dy} ]
    ]

PolarMap[ func_, {r0_, r1_, dr_}, {phi0_, phi1_, dphi_} ] :=
    Block[ {r, phi, coords},
        coords = Table[N[func[r Exp[I phi]]], {r, r0, r1, dr}, {phi, phi0, phi1, dphi}];
        Show[ MakeLines[coords], AspectRatio->Automatic, Axes->Automatic ]
    ]

PolarMap[ func_, {r0_:0, r1_}, {phi0_, phi1_} ] :=
    Block[{dr, dphi, plotpoints},
        plotpoints = PlotPoints /. Options[Plot3D];
        dr = (r1 - r0)/(plotpoints-1); dphi = (phi1 - phi0)/(plotpoints-1);
        PolarMap[ func, {r0, r1, dr}, {phi0, phi1, dphi} ]
    ]

End[(* "`Private`" *)]
EndPackage[]
```

ComplexMap2.m: Additional rules for default arguments

In[1]:= << ComplexMap2.m

The Riemann Zeta function with a default of 15 lines each.

In[2]:= **CartesianMap[Zeta, {0.1, 0.9}, {0, 20}]**

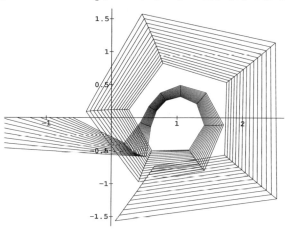

Fifteen lines usually give a poor picture. We can change the default by resetting the option PlotPoints for the function Plot3D[] to a larger value.

In[3]:= **SetOptions[Plot3D, PlotPoints->25];**

This new default value is now used and we get a better picture.

In[4]:= **CartesianMap[Zeta, {0.1, 0.9}, {0, 20}]**

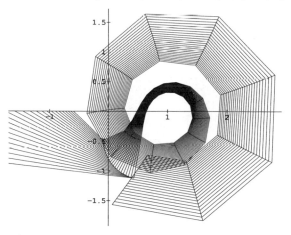

The square root function halves the angle of each complex number. The full circle from $-\pi$ to π is therefore mapped into half a circle. Subtracting a little bit from the starting point $-\pi$ avoids problems with the branch cut of the square root function. (Try it without it to see what happens.)

```
In[5]:= PolarMap[ Sqrt, {1}, {-Pi - 0.0001, Pi} ]
```

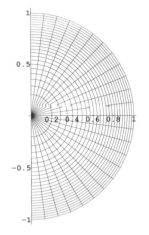

The example with the Zeta function shows that points that lie far away from the origin are clipped automatically. This is usually a desirable feature. We can, however, change the range of values plotted with the `PlotRange` option.

This picture shows more of the lines on the left, but details are lost overall.

```
In[6]:= Show[ %4, PlotRange -> {{-5, 3}, {-2, 2}} ]
```

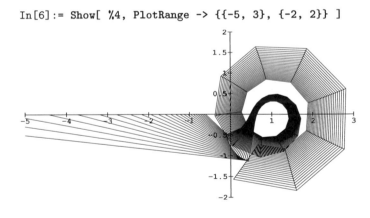

■ 1.5 Another Way to Add Defaults

In the preceding section, we added rules that allow us to leave out the increments in the range specifiers for `CartesianMap[]` and `PolarMap[]`. But what if we want to leave it out in one of the ranges and explicitly give an increment for the second one? We might want to say `PolarMap[Sqrt, {0, 1, 0.2}, {-Pi - 0.0001, Pi}]`. There is no rule which will match because the first rule for `CartesianMap[]` required both ranges to have three elements while the second one requires them to have two elements each (see the listing of **ComplexMap2.m**). We would need two more rules to cover the two mixed cases.

There is a way out. The only reason that we needed the extra rules was that the default values had to be computed and were not constants that could have been specified easily in the rule, as was the default value of 0 for the radius in `PolarMap[]`. The idea is to use some symbol as default value for the increment and then test for the presence of this symbol. The symbol `Automatic` can be used for this purpose. It is used for similar purposes as a value of many options of graphics functions. To test whether the value of the increment is this symbol, we use `SameQ[`e_1`, `e_2`]` (or e_1`===`e_2), which tests equality of symbols. The new rules that replace the old ones are the following:

```
CartesianMap[ func_, {x0_, x1_, dx_:Automatic}, {y0_, y1_, dy_:Automatic} ] :=
    Block[ {x, y, coords, plotpoints, ndx=dx, ndy=dy},
        plotpoints = PlotPoints /. Options[Plot3D];
        If[ dx === Automatic, ndx = N[(x1-x0)/(plotpoints-1)] ];
        If[ dy === Automatic, ndy = N[(y1-y0)/(plotpoints-1)] ];
        coords = Table[ N[func[x + I y]],
            {x, x0, x1, ndx}, {y, y0, y1, ndy} ];
        Show[ MakeLines[coords], AspectRatio->Automatic, Axes->Automatic ]
    ]

PolarMap[ func_, {r0_:0, r1_, dr_:Automatic}, {phi0_, phi1_, dphi_:Automatic} ] :=
    Block[ {r, phi, coords, plotpoints, ndr=dr, ndphi=dphi},
        plotpoints = PlotPoints /. Options[Plot3D];
        If[ dr === Automatic, ndr = N[(r1-r0)/(plotpoints-1)] ];
        If[ dphi === Automatic, ndphi = N[(phi1-phi0)/(plotpoints-1)] ];
        coords = Table[ N[func[r Exp[I phi]]],
            {r, r0, r1, ndr}, {phi, phi0, phi1, ndphi} ];
        Show[ MakeLines[coords], AspectRatio->Automatic, Axes->Automatic ]
    ]
```

part of **ComplexMap3.m**

As explained in section 2.5.10 of the *Mathematica* book, names for patterns, like `dx` and `dy` above, cannot be used as local variables inside the body of the rule. Therefore we declare two local variables `ndx` and `ndy` and initialize them with the values of `dx` and dy. We do not repeat the whole code for **ComplexMap3.m** at this point. The modifications introduced here will find their way into the next version in section 1.6.

■ 1.5.1 Defaults or Options?

Defaults are very convenient since they save a lot of repetitive typing. But too many of them can be confusing. Already, we have a possible conflict in the function `PolarMap[]`. There is a default for both the start value and the increment for the first range. So what does the range specifier {*a*, *b*} mean? Is *a* the start value, *b* the final value and the increment computed by default? Or does the start value default to 0 and is *a* the final value and *b* the increment? Matching from left to right seems more natural and indeed this is what happens.

In[1]:= << ComplexMap3.m

The identity function is a good test case since it does not deform the coordinate lines. As expected, the radius is between 1 and 2 and has the default number of increments (15 in this case). The increment for the angle is expressed as a fraction of π to get a nice-looking picture that closes up properly.

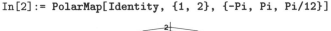

In[2]:= **PolarMap[Identity, {1, 2}, {-Pi, Pi, Pi/12}]**

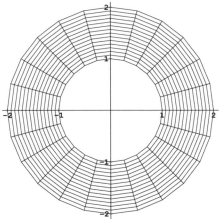

Arguments of a function whose meaning depends on their positions in the argument list are called *positional parameters*. The first three arguments of `CartesianMap[]` and `PolarMap[]` are positional. Because of their shortcomings, especially in interactive use, there are alternatives in some languages, the so-called *named arguments*. Named arguments are identified by their names and not by their positions in the argument list. They can be in any order. In *Mathematica* such named arguments are called *options*. Their use should be considered whenever your function has many features that should be user-settable, but when in most applications a default value is adequate. Graphics functions tend to have many such arguments as there are many aspects of a picture that you might want to change only occasionally. How to define options for your own functions is discussed further in section 3.3.

■ 1.6 Passing Options to Another Function

In the last example of section 1.4 we replotted the result of a prior call to `PolarMap[]` with an option changing the plot range. It would be useful if we could specify options to `Graphics[]` directly in a call to `CartesianMap[]` and `PolarMap[]`. These options would then simply be passed along to the `Show[]` command. The pattern that will match these options must allow for a variable number of them, including zero. This is done with a "triple blank", a pattern of the form `opts___`. The value of `opts` is not used inside `CartesianMap[]` or `PolarMap[]`, but is simply passed to the `Show[]` command. The modified code is below:

```
CartesianMap[ f_, {x0_, x1_, dx_:Automatic}, {y0_, y1_, dy_:Automatic}, opts___ ] :=
    Block[ {x, y, coords, plotpoints, ndx=dx, ndy=dy},
        plotpoints = PlotPoints /. Options[Plot3D];
        If[ dx === Automatic, ndx = N[(x1-x0)/(plotpoints-1)] ];
        If[ dy === Automatic, ndy = N[(y1-y0)/(plotpoints-1)] ];
        coords = Table[ N[f[x + I y]],
            {x, x0, x1, ndx}, {y, y0, y1, ndy} ];
        Show[ MakeLines[coords], opts, AspectRatio->Automatic, Axes->Automatic ]
    ]

PolarMap[ f_, {r0_:0, r1_, dr_:Automatic}, {phi0_, phi1_, dphi_:Automatic}, opts___ ] :=
    Block[ {r, phi, coords, plotpoints, ndr=dr, ndphi=dphi},
        plotpoints = PlotPoints /. Options[Plot3D];
        If[ dr === Automatic, ndr = N[(r1-r0)/(plotpoints-1)] ];
        If[ dphi === Automatic, ndphi = N[(phi1-phi0)/(plotpoints-1)] ];
        coords = Table[ N[f[r Exp[I phi]]],
            {r, r0, r1, ndr}, {phi, phi0, phi1, ndphi} ];
        Show[ MakeLines[coords], opts, AspectRatio->Automatic, Axes->Automatic ]
    ]
```

part of **ComplexMap4.m**

Note the placement of `opts` in the argument list of `Show[]`. It comes before the options `AspectRatio` and `Axes`. This allows you to override these values since the first encounter of an option is used. If we had put `opts` at the end there would have been no way to change the settings for `AspectRatio` or `Axes`.

In[1]:= << ComplexMap4.m

This is the same example as in subsection 1.5.1. We override the defaults for `AspectRatio` and `Axes` and also draw a frame around the picture.

In[2]:= PolarMap[Identity, {1, 2}, {-Pi, Pi, Pi/12},
            AspectRatio -> 0.5, Axes -> None, Framed -> True]

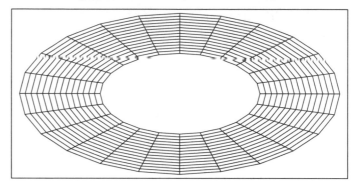

Here is another picture of the inversion (see section 1.3 for the first one). This time we use `CartesianMap[]` and also try to plot all points, even those far away from the center.

The images of the straight lines become circles under inversion. The coordinate lines that pass close by the origin are taken farthest away from it and don't look at all like circles there. Thus it is normally a good idea to cut off such points in the way that *Mathematica* does it, by default.

In[3]:= CartesianMap[1/Conjugate[#]&, {-2, 2, 4/19},
            {-2, 2, 4/19}, PlotRange->All]

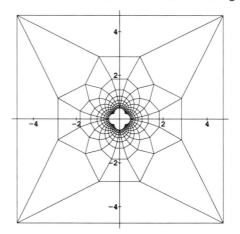

■ 1.7 Advanced Topic: Dealing with Singularities

So far, we have been careful to avoid examples where the function to be plotted had singularities at the grid points. Singularities are however an essential feature of complex functions. We have already encountered a function with a singularity: the inversion at the unit circle (the last example in section 1.3 on page 14). There we chose the plot ranges in such a way that we did not hit the origin, where the function is undefined (we would be dividing by zero there).

Very large values can also pose practical problems. Many PostScript devices cannot handle them and would render our picture wrongly. First we deal with the cases that are easy to repair. Then we will treat infinity and undefined function values.

■ 1.7.1 Huge Function Values

We have already seen examples where parts of the lines were cut off because they had endpoints with very large values. If the interesting values that we want to plot are all in a range between -10 and 10, say, then our picture will look the same whether some other values have an order of magnitude of 10^{100} or only 10^6. We can therefore simply replace such large values by smaller ones, provided that they lie in the same direction. The direction is given by the argument (or phase angle) of a complex number. We therefore replace every complex number whose absolute value is larger than some threshold by a number of the same argument but with a smaller absolute value. This can be done by a replacement rule inside the auxiliary function MakeLines[] in our package. The threshold is stored in a local variable. Along the same lines we also replace each number whose absolute value is very small by 0. Here is the new code for MakeLines[]:

```
huge = 10.0^6

MakeLines[points_] :=
    Block[{coords, lines, newpoints},
        newpoints = points /. z_?NumberQ :> huge z/Abs[z] /; Abs[z] > huge;
        newpoints = points /. z_?NumberQ :> 0.0 /; Abs[z] < 1/huge;
        coords = Map[ {Re[#], Im[#]}&, newpoints, {2} ];
        lines = Map[ Line, Join[coords, Transpose[coords]] ];
        Graphics[lines]
    ]
```

Dealing with large values in MakeLines[]

The quantity z/Abs[z] gives a complex number of magnitude 1 and the same argument as z. Multiplying it by the value huge gives the desired number in the same direction as z but with a smaller magnitude. We do this for all numbers that are larger than huge.

Some forms of infinity also have a direction associated with them. The most familiar is Infinity, which stands for $+\infty$. Internally all of these are represented by DirectedInfinity[z] where z is a (usually complex) number. Its argument is the direction of this infinity. To draw lines to these points, we can simply replace them with a finite number in the same direction. All three transformations can be collected in a single list of rules.

```
huge = 10.0^6

MakeLines[points_] :=
    Block[{coords, lines, newpoints},
        newpoints = points /.
            { z_?NumberQ :> huge z/Abs[z] /; Abs[z] > huge,
              z_?NumberQ :> 0.0 /; Abs[z] < 1/huge,
              DirectedInfinity[z_] :> huge z/Abs[z] };
        coords = Map[ {Re[#], Im[#]}&, newpoints, {2} ];
        lines = Map[ Line, Join[coords, Transpose[coords]] ];
        Graphics[lines]
    ]
```

Dealing with directed infinity in MakeLines[]

■ 1.7.2 Complex Infinity and Indeterminate Values

The expression 1/0 gives an infinite result without a direction associated with it. It is represented by DirectedInfinity[] without an argument. DirectedInfinity[] is normally printed as ComplexInfinity. The expression 0/0 is even undetermined in magnitude. It is represented by the symbol Indeterminate. There is no consistent way of representing these values in the plot of a function. If one of these two values occurs as a point in one of the lines, then the best we can do is to plot the line in two segments leaving out the bad point. This is considerably more tricky than anything we have done so far. We develop a function SplitLine[] that deals with a single list of numbers and returns a list of the parts between bad points. SplitLine[] is then applied to all the horizontal and vertical lines.

To make things more uniform, we replace each point that is not a number by the symbol Indeterminate. For each list of numbers that is to become a line in our picture, we use the function Position[] to obtain the position of all of its elements that are equal to Indeterminate. In a loop, we then pick out all subparts of the list between two occurrences of Indeterminate that are at least two apart, because a single good point between two bad ones does not give us a line segment. The beginning and end of the list can be treated uniformly by assuming a bad point at either end of the list. Here it is:

```
SplitLine[vl_] :=
    Block[{vll, pos, linelist = {}, low, high},
        vll = If[NumberQ[#], #, Indeterminate]& /@ vl;
        pos = Flatten[ Position[vll, Indeterminate] ];
        pos = Union[ pos, {0, Length[vll]+1} ];
        Do[ low = pos[[i]]+1;
            high = pos[[i+1]]-1;
            If[ low < high, AppendTo[linelist, Take[vll, {low, high}]] ],
            {i, 1, Length[pos]-1}];
        linelist
    ]
```

The function `SplitLine[]`

This is rather complicated and we present now a useful debugging technique to see how this function works. We execute the statements of this function one after the other with a small example as parameter. Instead of the parameter, we use a variable and assign it the value that we want to test. If a statement is deeply nested, it can be unwound and executed step by step. Let's try an example:

This list has two bad points in it. We assign it to the name of the pattern in the parameter list of `SplitLine[]`.

```
In[1]:= vl = {1.0, 2.0, ComplexInfinity,
                4.0, 5.0, 6.0, Indeterminate}
Out[1]= {1., 2., ComplexInfinity, 4., 5., 6.,
    Indeterminate}
```

We also need to execute any initializations in the local variable declarations of the block.

```
In[2]:= linelist = {}
Out[2]= {}
```

This maps all bad points into the symbol `Indeterminate`. Note that we leave out the terminating semicolon to see the result.

```
In[3]:= vll = If[NumberQ[#], #, Indeterminate]& /@ vl
Out[3]= {1., 2., Indeterminate, 4., 5., 6., Indeterminate}
```

The next statement is nested so we do it in two steps.

```
In[4]:= Position[vll, Indeterminate]
Out[4]= {{3}, {7}}
```

The percent sign comes in handy here.

```
In[5]:= pos = Flatten[%]
Out[5]= {3, 7}
```

This adds the two virtual bad points beyond the limits of the list.

```
In[6]:= pos = Union[ pos, {0, Length[vll]+1} ]
Out[6]= {0, 3, 7, 8}
```

We can execute the loop step by step by assigning the start value to the iterator variable explicitly.

```
In[7]:= i=1
Out[7]= 1
```

The variables `low` and `high` are used to get a good part of the list, one that lies between two bad points.

```
In[8]:= low = pos[[i]]+1
Out[8]= 1
```

```
In[9]:= high = pos[[i+1]]-1
Out[9]= 2
```

We can look at the value of the predicate in the `If[]` statement to see whether its second argument will be executed.

```
In[10]:= low < high
Out[10]= True
```

Again we unwind the statement. This picks out a good part of the list of points.

```
In[11]:= Take[vll, {low, high}]
Out[11]= {1., 2.}
```

This part is now appended to the result.

```
In[12]:= AppendTo[linelist, % ]
Out[12]= {{1., 2.}}
```

Once we are convinced that the loop does the right thing we can run through the remaining iterations. There is one catch here: we have to clear the value we assigned the iterator variable `i` first.

```
In[13]:= i=.
```

Note the changed start value for `i`, since we have already done the first iteration.

```
In[14]:= Do[ low = pos[[i]]+1; high = pos[[i+1]]-1;
           If[ low < high, AppendTo[linelist,
                               Take[vll, {low, high}]] ],
         {i, 2, Length[pos]-1}]
```

This would be the return value of the function `SplitLine[]`.

```
In[15]:= linelist
Out[15]= {{1., 2.}, {4., 5., 6.}}
```

If you are running *Mathematica* on a computer that allows you to cut and paste text between different windows on the screen, the above steps can be done quite easily. You simply keep the code of the function in a different window and transfer one line after the other to the *Mathematica* window.

Now we are ready to incorporate `SplitLine[]` into the code of `MakeLines[]`. We have to make one small change inside `MakeLines[]` in addition to mapping the function `SplitLine[]` to all the lines. The result of `SplitLine[]` is a list of lines. Collecting all of them together would give us lists of lists of lines. Applying `Flatten[`*list*`, 1]` gets rid of the extra level of lists. The code for the functions `CartesianMap[]` and `PolarMap[]` did not change at all. The advantages of local auxiliary procedures become apparent once more. Here is the final version of the function `MakeLines[]`:

```
huge = 10.0^6

MakeLines[points_] :=
    Block[{lines, newpoints},
        newpoints = points /.
            { z_?NumberQ :> huge z/Abs[z] /; Abs[z] > huge,
              z_?NumberQ :> 0.0 /; Abs[z] < 1/huge,
              DirectedInfinity[z_] :> huge z/Abs[z] };
        lines = Join[ newpoints, Transpose[newpoints] ];
        lines = Flatten[ SplitLine /@ lines, 1 ];
        lines = Map[ {Re[#], Im[#]}&, lines, {2} ];
        lines = Map[ Line, lines ];
        Graphics[lines]
    ]
```

The final version of `MakeLines[]`

These new versions of `SplitLine[]` and `MakeLines[]` replace the old ones and we call this version of our package **ComplexMap5.m**.

■ 1.7.3 A Thorough Test

We would like to have a function which exhibits all of the singularities and bad points that we have dealt with now. We can define one ourselves. It will basically be the identity, but at a few places will have the singularities treated in the previous two subsections.

```
f[z_] := Indeterminate /; Abs[z] < 10.^-3          (* Indeterminate near 0      *)
f[z_] := 10^100 /; Abs[z-2] < 10.^-3               (* very large near 2         *)
f[z_] := DirectedInfinity[-2-I] /; Abs[z-(-2+I)] < 10.^-3 (* infinite near -2+I        *)
f[z_] := DirectedInfinity[] /; Abs[z-(2-2I)] < 10.^-3    (* ComplexInfinity near 2-2I *)
f[z_] := z                                          (* identity elsewhere        *)
```

ComplexTest.m: A function with a few prescribed singularities

Read in the definition of f.

```
In[1]:= << ComplexTest.m
```

Read in the new version of **ComplexMap**.

```
In[2]:= << ComplexMap5.m
```

In the picture we see the line segments leading to 0 and to $2 - 2i$ missing. The lines leading to the points with function value 10^{100} and $(-2 - i)\infty$ point into the right direction.

```
In[3]:= CartesianMap[ f, {-2, 2, 1/3}, {-2, 2, 1/3},
            Axes->None ]
```

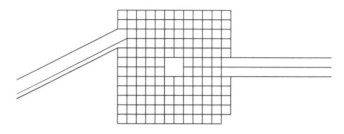

This is again the inversion at the unit circle. This time, however, we are not afraid of letting one of the grid points be the origin.

```
In[4]:= CartesianMap[ 1/Conjugate[#]&,
            {-2, 2, 0.2}, {-2, 2, 0.2} ]
```

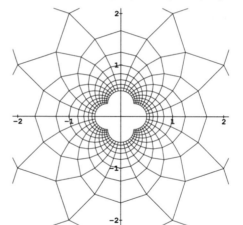

■ 1.8 Parameter Type Checking

So far we have not paid any attention to possibly bad parameters that a user of our package might accidentally type in. If a built-in function is called with a bad parameter value, it usually prints an error message and returns itself unevaluated.

What happens if `CartesianMap[]` or `PolarMap[]` is called with bad parameters? If the two ranges (the second and third arguments of these functions) are given with the wrong number of elements (only one or more than three), the rules will simply not match. Things are worse if the number of arguments is correct, but one of the values inside the range does not evaluate to a number. In this case the `Table[]` command inside the function body will generate an error message. Its value is then not a list of numbers and most of the following commands will also generate error messages. Finally, the `Show[]` command will complain that it has not received a valid graphics specification as input. If the user does not know how our function works internally (and there should be no reason for him to have to know), then these error messages will be very confusing because they are not recognized as a consequence of the original error. (Try for example to evaluate `CartesianMap[Log, { pi, pi, 0.1}, {-1, 1}]` with the common mistake of writing `pi` instead of `Pi`.)

To be "user-friendly," our program should check parameter values as well as it can. One obvious condition is that all the elements in the two ranges should evaluate to numbers. Here is `CartesianMap[]` with these checks added as conditions at the end:

```
CartesianMap[ func_, {x0_, x1_, dx_:Automatic}, {y0_, y1_, dy_:Automatic}, opts___ ] :=
    Block[ {x, y, points, plotpoints, ndx=dx, ndy=dy},
        ⋮
    ] /; NumberQ[N[x0]] && NumberQ[N[x1]] && NumberQ[N[y0]] && NumberQ[N[y1]]
```

Type checks in `CartesianMap[]`

It would be too restrictive to simply test for `NumberQ[x0]`. `x0` could be a constant (like `Pi`) or a function value (like `Sin[1]`) that is not a number but that does evaluate to a number when `N[]` is applied to it.

Another common mistake could be to call `CartesianMap[]` with extra arguments at the end that are not options of `Graphics3D[]`. We will later give a method of making sure that these options are all valid (see section 3.4), but for now we just want to check that these arguments are all syntactically correct, i.e., that they are rules of the form *name -> value*. Options are internally the same as replacement rules, so we require that their heads be the symbol `Rule`. This is easy to do. We change the parameter from `opts___` to `opts__Rule`. This is matched only by a sequence of rules.

So far we have not tried to check the types of the parameters dx and dy, which specify the increments. This case is different since they could default to Automatic which does not evaluate to a number. We must therefore allow that dx and dy either evaluate to numbers or be equal to the symbol Automatic.

```
CartesianMap[ func_, {x0_, x1_, dx_:Automatic}, {y0_, y1_, dy_:Automatic}, opts___ ] :=
    Block[ {x, y, points, plotpoints, ndx=dx, ndy=dy},
          ⋮
    ] /; NumberQ[N[x0]] && NumberQ[N[x1]] && NumberQ[N[y0]] && NumberQ[N[y1]] &&
         (NumberQ[N[dx]] || dx === Automatic) &&
         (NumberQ[N[dy]] || dy === Automatic)
```

More type checks in CartesianMap[ ]

Note that the the terms connected with the logical or (two vertical bars in *Mathematica* notation) have to be put in parentheses since the priority of && is higher than the priority of | |. These changes are part of **ComplexMap6.m**.

This completes Chapter 1. We will return to this example occasionally.

Chapter 2

Encapsulation Mechanisms for Packages

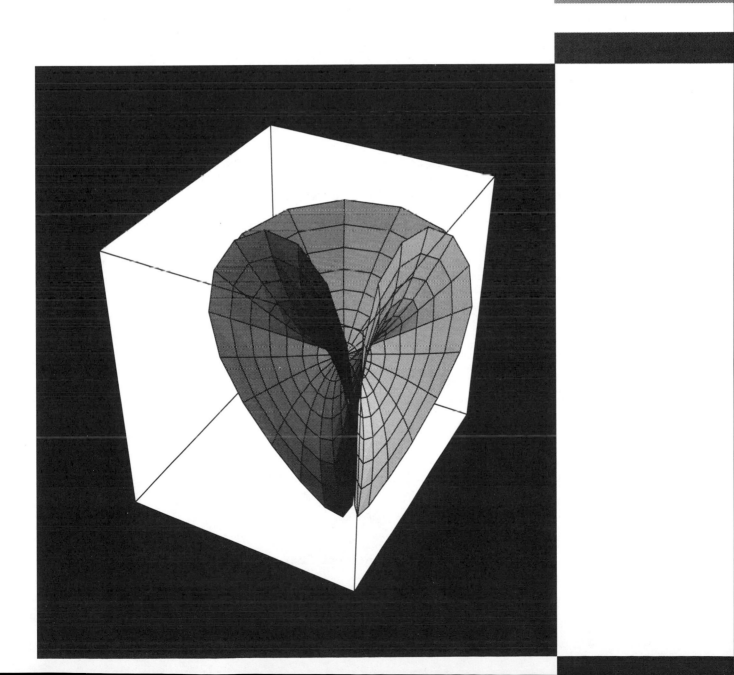

In Chapter 1 we have used some of the tools that *Mathematica* provides to write a package. Now we want to look in detail at all the facilities involving packages.

The first important concept is that of a *context*. Contexts provide a way of keeping variables in different packages separate. Another benefit of contexts is the ability to hide information from the user of our package. We want to keep local variables and auxiliary functions hidden from the user. Contexts are discussed in detail in section 2.1.

Section 2.2 introduces the concept of importing another package into a package. This allows us to use another package inside our own or to add to an already existing package.

In section 2.3 we present the tools for protecting a package against inadvertent modification by users. This is done in the same way as for built-in objects.

Next we present a *skeletal package*. It contains all the commands for setting up the proper contexts and can be used as a starting point for your own packages. Using such a template guarantees a certain uniformity in the overall appearance of a package, making it easier to understand packages that somebody else has written.

Finally we take a look at some special cases where the usual setup is inadequate. *Mathematica* gives you a great flexibility in choosing your programming style. It is however usually a good idea to limit yourself to certain standard mechanisms as they are explained in this chapter.

About the illustration overleaf:

A minimal surface. This one was generated as a parametric surface with coordinates

$$(r\cos\phi - \frac{r^2}{2}\cos 2\phi, -r\sin\phi - \frac{r^2}{2}\sin 2\phi, \frac{4}{3}r^{3/2}\cos\frac{3}{2}\phi).$$

Minimal surfaces have interesting mathematical properties. They are not easy to imagine given only their formulae. A picture helps a lot.

■ 2.1 All about Contexts

Recall from Chapter 1, section 1.2, that each symbol in *Mathematica* belongs to a context. There are two global variables inside *Mathematica* that control the creation of new symbols and the lookup of existing ones.

$Context	the current context
$ContextPath	the list of contexts to search

These two variables govern the lookup of contexts

When *Mathematica* encounters a symbol in the input that you type or that is read from a file, it searches the current context and then all the contexts on the context path for this symbol. If it cannot find one, a new symbol is created in the current context. See also sections 2.7.3 through 2.7.7 of the *Mathematica* book for additional explanations on contexts.

Normally, the values of the variables $Context or $ContextPath should not be changed directly. It is better to use the commands provided for context manipulation. They make the necessary changes to these variables and also perform error checking.

BeginPackage["*Context*`"]	start a package
EndPackage[ ]	end a package
Begin["*Context*`"]	change the current context
End[ ]	return to the previous context
EndAdd[ ]	return to the previous context and also add the current context to the search path

Commands to manipulate contexts

To see how these commands change the values of $Context and $ContextPath, let us look in detail at what happens when a package is read in. We simulate reading in a stripped-down version of the example from Chapter 1. We take out all the actual code and are left with just a few declarations:

```
BeginPackage["ComplexMap`"]

CartesianMap::usage = "CartesianMap[f,...] plots a map."

Begin["`Private`"]

huge = 10.0^6

MakeLines[points_] :=
    Block[{lines, newpoints},
              ⋮
    ]

CartesianMap[ f_,... ] := ...

End[ ]
EndPackage[ ]
```

An excerpt from **ComplexMap.m**

In the following table, we show the effect of each line in the package on the values of the variables $Context and $ContextPath. The entries are filled in only if the command changed them. We also show the fully qualified names of the symbols encountered in the input. The first line is not part of the package. It simulates some computations that we did before reading in the package. At the end, we return to this global context and use the function CartesianMap[] that we have just defined.

	Command	Symbols	$Context	$ContextPath
1	Factor[y^2-1]	System`Factor, Global`y	Global`	{Global`, System`}
2	<<ComplexMap.m			
	the following lines are read from the package			
3	BeginPackage["ComplexMap`"]	System`BeginPackage	ComplexMap`	{ComplexMap`, System`}
4	CartesianMap::usage = "..."	ComplexMap`CartesianMap		
5	Begin["`Private`"]	System`Begin	ComplexMap`Private`	
6	huge = 10.0^6	ComplexMap`Private`huge		
7	MakeLines[points_] :=	ComplexMap`Private`MakeLines		
		ComplexMap`Private`points		
8	Block[{lines,...},...]	ComplexMap`Private`lines		
9	CartesianMap[f_,...] :=	ComplexMap`CartesianMap		
		ComplexMap`Private`f		
10	End[ ]		ComplexMap`	
11	EndPackage[ ]		Global`	{ComplexMap`, Global`, System`}
12	CartesianMap[Sin,...]	ComplexMap`CartesianMap		
		System`Sin		
13	MakeLines[...]	Global`MakeLines		

Here is a line-by-line discussion of what happens exactly in the example above:

1. This line shows a typical calculation that could have been done just before reading in the package **ComplexMap.m**. The values of the variables $Context and $ContextPath are the default ones when *Mathematica* starts up.

2. Here we give the command to read in the package. The following lines are the contents of the package; they are not typed in by the user.

3. BeginPackage[] sets the value of $Context. $ContextPath always becomes {*PackageContext*`, System`} independent of what it was before. Note especially that Global` is not on the context search path. In a package there is therefore no danger of wrongly accessing any objects that have been defined in the *Mathematica* session so far. The package always starts in a clean state.

4. Defining a usage message for CartesianMap creates the symbol since it is not built in and does not exist in the current context. It is, therefore, created in the current context, which is the package context ComplexMap`.

5. The command Begin["*Context*`"] changes the current context. It does not affect the context path. The argument "`Private`" begins with a context mark ` and the context is, therefore, a subcontext of the current context, with full name ComplexMap`Private`. The consequence is that newly created symbols will be put in the context ComplexMap`Private`.

6. The variable huge is local to the package. Its value is, however, preserved from one call of CartesianMap[] to the next, since it is not declared as a local variable to CartesianMap[]. Such variables are sometimes called *static variables*.

7. The command MakeLines[] too is local to the package.

8. The variable lines is local to the command MakeLines[]. Its value is not preserved across calls.

9. Finally, here is the definition of the function CartesianMap[]. The symbol CartesianMap has already been created (in line 4) in the context ComplexMap`. Since this context is on the context path the symbol is found there and the definition is for the existing symbol.

10. The command End[] undoes the previous Begin[] and restores the current context to ComplexMap`. The context ComplexMap`Private` is not on the context path, and any symbols defined in this context are therefore no longer accessible.

11. EndPackage[] restores the current context to what it was before the command BeginPackage[]. The context path is also restored, but the package context is added in front of it.

12. Now we are back in our interactive *Mathematica* session and can use the command CartesianMap[] since the context in which it was defined appears in the context path.

13. The command `MakeLines[]`, however, cannot be used. The symbol is not found
    in its context `ComplexMap`Private``. A new symbol is created in the global
    context that has nothing to do with the other one.

The net effect of reading in a package is to add a new context in front of the context
path and to define some functions in this context. Functions that are made available in
a self-contained program unit for use outside of it are said to be *exported* from it. The
package controls which of its functions it wants to export. This mechanism of explicitly
exporting objects from a program unit is found in one form or another in most modern
programming languages. It is an important software engineering tool.

Context names themselves are not symbols and must be quoted when used as arguments of a command. A symbol in any context can be specified with its fully qualified
name as *Context`symbol*. It is therefore not entirely true that the auxiliary function
`MakeLines[]` is inaccessible. By using `ComplexMap`Private`MakeLines`, we could
access it even though the context in which it is defined is not on the search path. Needless
to say, this practice is strongly discouraged.

■ 2.2 Packages that Use Other Packages

Recall from section 2.1 that the command BeginPackage["*Context*`"] resets the context path to {*Context*`, System`}, independent of its former value. One consequence is that symbols that the user of your package has defined before reading in the package do not get in the way since such symbols are defined in the global context. But it also means that any other packages that have been read in are not accessible inside your package. You might want to use a command defined in another package inside your own package, however.

■ 2.2.1 Importing Another Package

BeginPackage[] has optional arguments that specify contexts that are to be left on the search path. If they are not already on the search path, then they are first read in using the command Needs["*Context*`"] implicitly. An example of such a package is the standard package **Graphics/Shapes.m**. It defines a command to rotate a three-dimensional graphics object and uses the standard package **Geometry/Rotations.m** to compute the rotation matrix. Here is a small excerpt showing how this is set up:

```
BeginPackage["Shapes`", "Geometry`Rotations`"]

RotateShape::usage = "..."

Begin["`Private`"]

RotateShape[ shape_, phi_, theta_, psi_ ] :=
    Block[{rotmat = RotationMatrix3D[N[phi], N[theta], N[psi]]},
              ⋮
    ]

End[ ]
EndPackage[ ]
```

Part of **Shapes.m**

The function RotationMatrix3D[] used inside RotateShape[] comes from the package **Geometry/Rotations.m**. Here is the detailed analysis of the context changes in the same format as in section 2.1.

	Command	Symbols	$Context	$ContextPath
1	BeginPackage["Shapes`", "Geometry`Rotations`"]		Shapes`	{Shapes`, Geometry`Rotations`, System`}
2	RotateShape::usage = "..."	Shapes`RotateShape		
3	Begin["`Private`"]		Shapes`Private`	
4	RotateShape[shape_,...] :=	Shapes`RotateShape Shapes`Private`shape		
5	Block[{rotmat= RotationMatrix3D[..]},..]	Shapes`Private`rotmat Geometry`Rotations`RotationMatrix3D		
6	End[ ]		Shapes`	
7	EndPackage[ ]		Global`	{Shapes`, Geometry`Rotations`, Global`, System`}

As with the command `Needs[]` itself, no package is read in if the context given as an optional argument to `BeginPackage[]` is already on the search path. A package is therefore read in only once even if it is used inside several packages that you read in your session one after the other.

The optional arguments of `BeginPackage[]` specify those packages that are needed inside your package. Such packages are said to be *imported* in your package. Mentioning imported packages at the start of your package is also an important element of documentation. It makes all dependencies on other packages explicit. Stating exactly what your package depends on is another important principle of software design.

■ 2.2.2 Hidden Import

Any package read in through the mechanism just explained is left on the search path after your package has been read in. It is therefore also made available to the user of your package. This is sometimes desirable or does not matter very much. But it could also hide other functions that the user of your package has defined or read in before your package. The user should not have to be concerned about other packages that are made available as a consequence of reading in yours. There is a way of making another package available to your own without leaving it on the search path after the end of your package. Instead of mentioning the package as as an optional argument of `BeginPackage[]`, you can also read it in with a `Needs[]` command inside your package, after the call to `BeginPackage[]`. That your package imports this other package is then hidden from the user, and this method is therefore termed *hidden import*. Here is the modified code fragment of **Shapes.m** followed by the detailed analysis of the context changes.

```
BeginPackage["Shapes`"]

Needs["Geometry`Rotations`"]

RotateShape::usage = "..."

Begin["`Private`"]
```

```
RotateShape[ shape_Graphics3D, phi_, theta_, psi_ ] :=
    Block[{rotmat = RotationMatrix3D[N[phi], N[theta], N[psi]]},
            ⋮
    ]
End[ ]
EndPackage[ ]
```

A modified version of **Shapes.m**

	Command	Symbols	$Context	$ContextPath
1	BeginPackage["Shapes`"]		Shapes`	{Shapes`, System`}
2	Needs["Geometry`Rotations`"]			{Geometry`Rotations`, Shapes`, System`}
3	RotateShape::usage = "..."	Shapes`RotateShape		
4	Begin["`Private`"]		Shapes`Private`	
5	RotateShape[shape_,...] :=	Shapes`RotateShape Shapes`Private`shape		
6	Block[{rotmat= RotationMatrix3D[..]},..]	Shapes`Private`rotmat Geometry`Rotations`RotationMatrix3D		
7	End[ ]		Shapes`	
8	EndPackage[ ]		Global`	{Shapes`, Global`, System`}

After reading in this version of **Shapes.m** the context Rotations` is not on the search path and cannot be accessed by the user. Note that inside **Shapes.m** the order of the contexts Shapes` and Geometry`Rotations` on the search path is reversed. This poses no problems since the current context is still either Shapes` or Shapes`Private` and so new symbols like RotateShape are still created in the correct context and not in Geometry`Rotations`.

There is one potential problem with this hidden import of packages. If another package also tries to import a package that has already been read in somewhere else, then it would be read in twice. Many packages are not designed to be read in more than once. If packages are requested only through the command BeginPackage[] or through Needs[] *before* the command BeginPackage[] and not through hidden import or the explicit command <<*Package*, this cannot happen. Hidden import is therefore best used only for packages that were explicitly designed to be used that way, for example auxiliary packages for one of your own packages.

■ 2.2.3 Context Names and Package Names

When a context name is specified in the Needs[] command or as one of the optional arguments of BeginPackage[], *Mathematica* has to derive the name of a file to read in. By convention, the name of the file is of the form *Context*.m where *Context* is the context name without the final `. If the context name is composed of several parts separated by context marks, then these context marks are translated into appropriate path separators for your file system (/ under UNIX for example). The command `Needs["Geometry`Rotations`"]` would try to read in the package **Geometry/Rotations.m**. This is the usual case since all standard packages are put into subdirectories of the Package directory.

`Needs[]` has an optional second argument that allows you to specify a different file name. This, however, is not possible in the `BeginPackage[]` command. One way around it is to precede `BeginPackage[]` by a call to `Needs[]`. If the package context `Rotations`` were in the file **TestRotations.m** instead of **Rotations.m**, then the following is a possible way to import this package into **Shapes.m** in the same way as was done in subsection 2.2.1:

```
Needs["Rotations`", "TestRotations.m"]
BeginPackage["Shapes`", "Rotations`"]
      ⋮
EndPackage[ ]
```

Another variant of **Shapes.m**

■ 2.2.4 Extending other Packages

You can look at importing another package in a different way than we did in subsection 2.2.1. Instead of merely using one of the functions in the imported package, you could also think of adding some more functions to the ones defined in the imported package since the imported package will be available to the user of your package (remember that it remains on the context search path). Here is an example.

The package **Graphics/ParametricPlot3D.m** contains a function `SphericalPlot3D[]` for making plots in spherical coordinates. Let us now add a function for making plots in *cylindrical* coordinates. The package **CylindricalPlot3D.m** effectively adds the function `CylindricalPlot3D[]` to the collection of functions defined in the standard package **ParametricPlot3D.m**. In `ParametricPlot3D[]`, you specify the x, y, and z coordinates of points as a function of two variables to generate a surface. With `SphericalPlot3D[]` you specify the radius r of points in terms of the two angles θ and ϕ. In the new function `CylindricalPlot3D[]`, you specify z in terms of r and ϕ. r measures the distance from the origin to the point in the x–y plane directly under the corresponding point of the surface and ϕ measures the angle from the x axis (This is not the same r as in spherical coordinates). ϕ ranges from 0 to 2π.

```
BeginPackage["CylindricalPlot3D`", "Graphics`ParametricPlot3D`"]

CylindricalPlot3D::usage = "CylindricalPlot3D[z, {r-range}, {phi-range}, (options...)]
    plots z as a function of r and phi."

Begin["`Private`"]

CylindricalPlot3D[ z_, rlist:{r_, __}, plist:{phi_, __}, opts___ ] :=
        ParametricPlot3D[{r Cos[phi], r Sin[phi], z}, rlist, plist, opts] /;
            3 <= Length[rlist] <= 4 && 3 <= Length[plist] <= 4

End[]
EndPackage[]
```

CylindricalPlot3D.m

CylindricalPlot3D[] works by converting the cylindrical coordinates into Cartesian coordinates and then simply calling the function ParametricPlot3D[]. We do not even need a local variable so no Block[] statement is necessary.

Note the form of the patterns that we use for the two ranges. We need the range as a whole to pass it on to ParametricPlot3D[], so we give it a name with rlist:*pattern*. We also need the name of the variable which is the first entry in the range, so we give it a name as well. We are, however, not interested in the other elements in the range (the start and end values and possibly the increment), so we use a pattern that matches the additional elements without giving them a name. The "double blank" __ does just that. The condition at the end makes sure that the lists have the correct length—Three elements for a default increment (which is handled inside ParametricPlot3D[]) or four elements if the increment is specified explicitly. Any additional arguments are simply passed along. ParametricPlot3D[] uses the same mechanism for specifying options to the Show[] command as does our example ComplexMap.m from Chapter 1.

In[1]:= << CylindricalPlot3D.m

The context path shows that not only **CylindricalPlot3D.m** has been read, but also **Graphics/ParametricPlot3D.m**.

```
In[2]:= $ContextPath
Out[2]= {CylindricalPlot3D`, Graphics`ParametricPlot3D`,
    Global`, System`}
```

In this example, z depends only on r and the picture is therefore rotationally symmetric around the vertical axis. What we see here is a parabola rotated around the vertical axis.

```
In[3]:= CylindricalPlot3D[ r^2, {r, 0, 1}, {phi, 0, 2Pi} ]
```

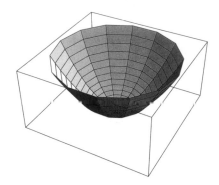

By specifying less than the full range of $(0, 2\pi)$ for ϕ we get a cut out view which sometimes helps in visualization. We can also get rid of the box and change the viewpoint.

```
In[4]:= CylindricalPlot3D[ r^2,
            {r, 0, 1},{phi, -Pi/4, 5Pi/4},
            Boxed->False, ViewPoint->{1.3, -2.4, 1.6} ]
```

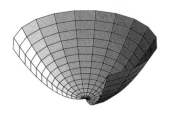

The functions from the imported package **ParametricPlot3D.m** are also available.

```
In[5]:= SphericalPlot3D[ Sin[theta]^2,
            {theta, 0, Pi}, {phi, 0, 3Pi/2} ]
```

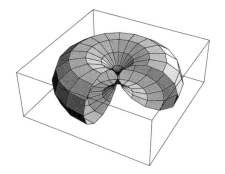

■ 2.3 Protection of Symbols in a Package

■ 2.3.1 Protecting Exported Symbols

Built-in commands are protected against accidental modification by a user. The commands `Protect[]` and `Unprotect[]` are discussed in section 2.2.8 of the *Mathematica* book. The commands defined in a package can be protected in the same way and will then behave just like built-in ones. All you have to do is to give the command `Protect[`*symb₁*`,` *symb₂*`,...]` at the end of the package (between `End[]` and `EndPackage[]`), where the *symbᵢ*'s are all the symbols to be exported from the package. You do not have to protect auxiliary functions defined in the private context since they cannot be accessed by the user anyway. Here is a version of the example **ComplexMap.m** from Chapter 1 with added protection:

```
BeginPackage["ComplexMap`"]

CartesianMap::usage = "..."
PolarMap::usage = "..."

Begin["`Private`"]
        ⋮
End[ ]

Protect[CartesianMap, PolarMap]

EndPackage[]
```

Protecting symbols in **ComplexMap.m**

A package with protected symbols cannot be read in twice. The second time around the symbols would already be protected when the rules and usage messages are defined, giving you error messages. Therefore, you usually do not include the `Protect[]` command while a package is still under development. On the other hand, you might want to put in a statement `Clear[`*symb₁*`,` *symb₂*`,...]` near the beginning of the package. This makes sure that all old rules are forgotten when you modify the rules and read in the modified package into the same *Mathematica* session many times during debugging. After having found the next-to-last bug in your code, you then insert the protection commands and remove the `Clear[]` at the beginning. (The *last* bug in any piece of code is, of course, invariably found by the first user of the code and never by the programmer.)

The protection of these symbols can still be removed by using `Unprotect[]`, just as for built-in symbols. If you really want to disable this, you can lock the symbols in addition to protecting them. The attributes of a locked symbol cannot be changed at all so this action is final. In the above example you could add the command

SetAttributes[{CartesianMap, PolarMap}, Locked]

after the call to Protect[].

■ 2.3.2 Unprotecting System Symbols

A complementary case arises if a package defines additional rules for built-in functions. In this case, the affected symbols have to be unprotected at the beginning of the package and should be protected again at the end.

Here is an example package **ReIm.m** that defines additional rules for the built-in functions Re[] and Im[] that allow simplification of symbolic expressions involving them. In *Mathematica*, variables that do not have a value are treated as standing for any quantity, including complex numbers. The expression Re[x] does, therefore, not simplify to x. In some applications, you assume that variables stand for real numbers only and then you would like to simplify Re[x] to x.

To declare a variable *var* to be real, set its imaginary part to 0 with the command *var*/: Im[*var*] = 0. Rules for Re[] and Im[] defined in **ReIm.m** will then perform the simplifications. Here are a few easy simplification rules for addition and multiplication:

```
BeginPackage["ReIm`"]

Begin["`Private`"]

Unprotect[Re, Im]

Re[x_] := x  /; Im[x] == 0

Re[x_+y_] := Re[x] + Re[y]
Im[x_+y_] := Im[x] + Im[y]

Re[x_ y_] := Re[x] Re[y] - Im[x] Im[y]
Im[x_ y_] := Re[x] Im[y] + Im[x] Re[y]

Protect[Re, Im]

End[ ]

EndPackage[ ]
```

The first version of **ReIm.m**

	In[1]:= << ReIm1.m

Nothing is known about x. It does not
simplify.

```
In[2]:= Re[x]
Out[2]= Re[x]
```

a is declared to be real. This declaration has to
be associated with a since Im is protected.

```
In[3]:= a/: Im[a] = 0;
```

It does what we expect.

```
In[4]:= Re[a]
Out[4]= a
```

This is partially simplified using the knowledge
we have about a.

```
In[5]:= Im[a x]
Out[5]= a Im[x]
```

One feature of **ReIm.m** is not quite the way we want it. If the user had already
unprotected Re, for example, then reading in our package would protect it again! This
should not happen and there is a way around it. The function Unprotect[] returns as
its value a list of all those of its arguments that it did actually unprotect. Arguments that
were already unprotected will be skipped. Instead of protecting all the symbols again at
the end, we only protect those that were returned by Unprotect[].

```
BeginPackage["ReIm`"]

Begin["`Private`"]

protected = Unprotect[Re, Im]

Re[x_] := x   /; Im[x] == 0

Re[x_+y_] := Re[x] + Re[y]
Im[x_+y_] := Im[x] + Im[y]

Re[x_ y_] := Re[x] Re[y] - Im[x] Im[y]
Im[x_ y_] := Re[x] Im[y] + Im[x] Re[y]

Protect[ Release[protected] ]

End[ ]

EndPackage[ ]
```

A better version of **ReIm.m**

The return value of the Unprotect[] command is saved in a local variable. Its
value is then used in the command Protect[] and only those symbols are then protected. Protect[] does not evaluate its arguments because it operates on the symbols
themselves and not on their values. It is thus necessary to force evaluation with the
command Release[], otherwise the symbol protected itself would be protected.

Here we unprotect the built-in symbol Re.

```
In[1]:= Unprotect[Re]

Out[1]= {Re}
```

Re no longer has the attribute Protected.

```
In[2]:= Attributes[Re]

Out[2]= {Listable}
```

Im is still protected.

```
In[3]:= Attributes[Im]

Out[3]= {Listable, Protected}
```

Now let's read in the package and see what happens to the protection of Re and Im.

```
In[4]:= << ReIm2.m
```

Both are as before.

```
In[5]:= Attributes[Re]

Out[5]= {Listable}
```

```
In[6]:= Attributes[Im]

Out[6]= {Listable, Protected}
```

In Appendix B, you will find the complete listing of **ReIm.m**, an improved version of the standard package **Algebra/ReIm.m**. It contains many more rules for simplification of Re[], Im[] and Conjugate[].

■ 2.4 A Skeletal Package

If you look at the examples presented so far and at other packages, you will notice that the framework of a package is mostly the same for all of them. We have developed this framework in Chapter 1 and the preceding sections of this chapter. We are now ready to collect everything together in a template or skeleton for packages. (As with real skeletons, the flesh is missing and has to be provided by the author of the package.) The cornerstones of every package are the context manipulation commands. Then come the usage messages for the functions that are to be exported, then the details about imports of other packages and the protection of symbols.

The package **Skeleton.m** is syntactically correct so it can be read into *Mathematica*. To avoid error messages, you need to provide the three files **Package1.m**, **Package2.m**, and **Package3.m** as well. They need to contain only a few lines that define the contexts.

```
BeginPackage["Package1`"]
EndPackage[]
```

Package1.m: One of the imported packages

If you are writing your own package, you can take **Skeleton.m** as a starting point. Change all the names of the functions and the package itself (including the context name in BeginPackage[]). You should also make sure to delete any of the features that you do not use, for example, the statements for importing other packages or for defining rules for built-in objects. As explained in subsection 2.3.1, you might want to comment out the statement to protect the exported symbols while your package is still being debugged.

```
(* Skeleton.m -- a skeletal package *)
(* set up the package context, included any imports *)
BeginPackage["Skeleton`", "Package1`", "Package2`"]
Needs["Package3`"]    (* read in any hidden imports *)

(* usage messages for the exported functions and the context itself *)
Skeleton::usage = "Skeleton.m is a package that does nothing."
Function1::usage = "Function1[n] does nothing."
Function2::usage = "Function2[n, (m:17)] does even more nothing."

Begin["`Private`"]    (* begin the private context *)

(* unprotect any system functions for which rules will be defined *)
protected = Unprotect[ Sin, Cos ]

(* definition of auxiliary functions and local (static) variables *)
Aux[f_] := Do[something]
staticvar = 0

(* error messages for the exported objects *)
Skeleton::badarg = "You twit, you called `1` with argument `2`!"

(* definition of the exported functions *)
Function1[n_] := n
Function2[n_, m_:17] := n m /; n < 5 || Message[Skeleton::badarg, Function2, n]

(* rules for system functions *)
Sin/: Sin[x_]^2 := 1 - Cos[x]^2

Protect[ Release[protected] ]     (* restore protection of system symbols *)
End[]          (* end the private context *)
Protect[ Function1, Function2 ]    (* protect exported symbols *)
EndPackage[]  (* end the package context *)
```

Skeleton.m: A template for packages

■ 2.5 Special Cases

Sometimes the usual setup for a package is not adequate. In this section we present two cases that deviate from the usual form of a package as given in section 2.4.

■ 2.5.1 Importing the Global Context

Putting an exported symbol in a separate context has one problem if that symbol already exists in the global context. Here is a *Mathematica* session that illustrates the problem.

We try to use a function, but forget to read in the package first. *Mathematica* does not know about this function, so it returns your input. But it has created the symbol `CylindricalPlot3D` in the global context.

```
In[1]:= CylindricalPlot3D[ 1.5 Sqrt[1 + r^2],
                           {r, 0, 2}, {phi, 0, 2Pi} ]

                                            2
Out[1]= CylindricalPlot3D[1.5 Sqrt[1 + r ], {r, 0, 2},
          {phi, 0, 2 Pi}]
```

We want to correct the mistake and read in the package. (This package is described in subsection 2.2.4.)

```
In[2]:= << CylindricalPlot3D.m
```

Now let's try again. It still does not work!

```
In[3]:= CylindricalPlot3D[ 1.5 Sqrt[1 + r^2],
                           {r, 0, 2}, {phi, 0, 2Pi} ]

                                            2
Out[3]= CylindricalPlot3D[1.5 Sqrt[1 + r ], {r, 0, 2},
          {phi, 0, 2 Pi}]
```

The symbol `CylindricalPlot3D` in the global context is found first and hides the one that was defined in **CylindricalPlot3D.m**.

```
In[4]:= Context[CylindricalPlot3D]
Out[4]= Global`
```

This command asks for all symbols with name `CylindricalPlot3D` in all contexts. There are two of them indeed. (The second one is not printed as `Global`CylindricalPlot3D` because it can be accessed without typing its context.)

```
In[5]:= ?*`CylindricalPlot3D

CylindricalPlot3D`CylindricalPlot3D
CylindricalPlot3D
```

You could use the function by always typing it as `CylindricalPlot3D`CylindricalPlot3D` but that would be awkward. A drastic action is to remove the offending symbol from the symbol table.

```
In[5]:= Remove[CylindricalPlot3D]
```

Now it finds the correct one since the symbol in the global context no longer exists, and we finally get our hyperboloid.

```
In[6]:= CylindricalPlot3D[ 1.5 Sqrt[1 + r^2],
                           {r, 0, 2}, {phi, 0, 2Pi} ]
```

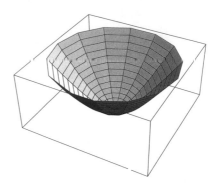

In cases where we anticipate that this problem may occur frequently, there is a nice trick to avoid it. Simply mention `Global`` as one of the contexts to be imported into your package! No file will be read since this context already exists. This idea is used in the standard package **Trigonometry.m** since it is so fundamental that we anticipated that users would frequently fall into the trap shown above. Here is its outline:

```
BeginPackage["Trigonometry`", "Global`"]

TrigCanonical::usage = "TrigCanonical[expr] applies basic trigonometric
    simplifications to expr (e.g. Sin[-x] --> -Sin[x])."

Begin["`Private`"]

TrigCanonicalRel = {
    Sin[n_?Negative x_ + y_] :> -Sin[-n x - y] /; Order[x, y] == 1 && NumberQ[n],
        ⋮
    }

TrigCanonical[e_] := e //. TrigCanonicalRel

End[]

Protect[ TrigCanonical, ... ]

EndPackage[]
```

The Outline of **Trigonometry.m**

Now we try to recreate the same circumstances as at the beginning of this section.

Again we use the function before it is defined. This creates the symbol `TrigCanonical` which is also exported from **Trigonometry.m**.	`In[1]:= TrigCanonical[Sin[-2a + x]]`
	
	`Out[1]= TrigCanonical[Sin[-2 a + x]]`

Now we read in the package.	`In[2]:= << Trigonometry.m`

But this time there is only one symbol.	`In[3]:= ?*'TrigCanonical`
	
	`TrigCanonical[expr] applies basic trigonometric`
	`    simplifications to expr (e.g. Sin[-x] —> -Sin[x]).`

It is the one we created in input line 1.	`In[3]:= Context[TrigCanonical]`
	
	`Out[3]= Global'`

But the rules from **Trigonometry.m** do work.	`In[4]:= TrigCanonical[Sin[-2a + x]]`
	
	`Out[4]= -Sin[2 a - x]`

Here is a table that traces the values of `$Context` and `$ContextPath` and the names of the symbols during the *Mathematica* session above.

	Command	Symbols	$Context	$ContextPath
1	`TrigCanonical[Sin[-2a + x]]`	`Global'TrigCanonical`	`Global'`	`{Global', System'}`
2	`<<Trigonometry.m`			
	the following lines are read from the package			
3	`BeginPackage["Trigonometry'", "Global'"]`		`Trigonometry'`	`{Trigonometry',` `Global', System'}`
4	`TrigCanonical::usage = "..."`	`Global'TrigCanonical`		
5	`Begin["'Private'"]`		`Trigonometry'Private'`	
6	`TrigCanonicalRel = { ... }`	`Trigonometry'Private'TrigCanonicalRel`		
7	`TrigCanonical[e_] := ...`	`Global'TrigCanonical`		
8	`End[ ]`		`Trigonometry'`	
9	`EndPackage[ ]`		`Global'`	`{Trigonometry',` `Global', System'}`
	here we return to the top level			
10	`TrigCanonical[Sin[-2a + x]]`	`Global'TrigCanonical`		

Inside the package **Trigonometry.m** the context `Global'` is left on the context path and so all the rules defined for `TrigCanonical` are attached to the already existing symbol `Global'TrigCanonical`. No new symbol in the package context is created. If the symbol `TrigCanonical` had not been created before reading in the package (the normal case) then it would have been created as usual.

■ 2.5.2 Tiny Packages

If we write code for one or two small commands that do not have any auxiliary functions, it is probably not worth creating a full-blown package. Nevertheless, it is a good idea to put the definitions into a separate, private context to avoid the kind of problems outlined at the beginning of section 1.2. Such a mini-package will look like this:

```
ExpandIt::usage = "ExpandIt[e] expands all numerators and denominators in e "

Begin["`Private`"]

ExpandIt[x_Plus] := ExpandIt /@ x
ExpandIt[x_] := Expand[ Numerator[x] ] / Expand[ Denominator[x] ]

End[]
Null
```

ExpandIt.m: A tiny package

The context used for the implementation of `ExpandIt[]` is specified as a subcontext of the current context—whatever it is at the time the tiny package is read. Commands like this can be put into the initialization file **init.m** and are then available in every *Mathematica* session you start.

$$In[1] := \text{<< ExpandIt.m}$$

Here is an expression with numerators and denominators that you might want to expand separately.

$$In[2] := \text{(1 + Sqrt[5])}\wedge 2/3 + \text{(a + b)}\wedge 2/\text{(e - I)}\wedge 3$$

$$Out[2] = \frac{(1 + Sqrt[5])^2}{3} + \frac{(a + b)^2}{(-I + e)^3}$$

Each numerator and denominator is expanded.

$$In[3] := \text{ExpandIt[\%]}$$

$$Out[3] = \frac{6 + 2\ Sqrt[5]}{3} + \frac{a^2 + 2\ a\ b + b^2}{I - 3\ e + -3\ I\ e^2 + e^3}$$

Compare this with the built-in `ExpandAll[]` that puts each term of the numerator over a separate copy of the denominator.

$$In[4] := \text{ExpandAll[\%\%]}$$

$$Out[4] = 2 + \frac{2\ Sqrt[5]}{3} + \frac{a^2}{I - 3\ e + -3\ I\ e^2 + e^3} + $$

$$\frac{2\ a\ b}{I - 3\ e + -3\ I\ e^2 + e^3} + \frac{b^2}{I - 3\ e + -3\ I\ e^2 + e^3}$$

Defaults and Options

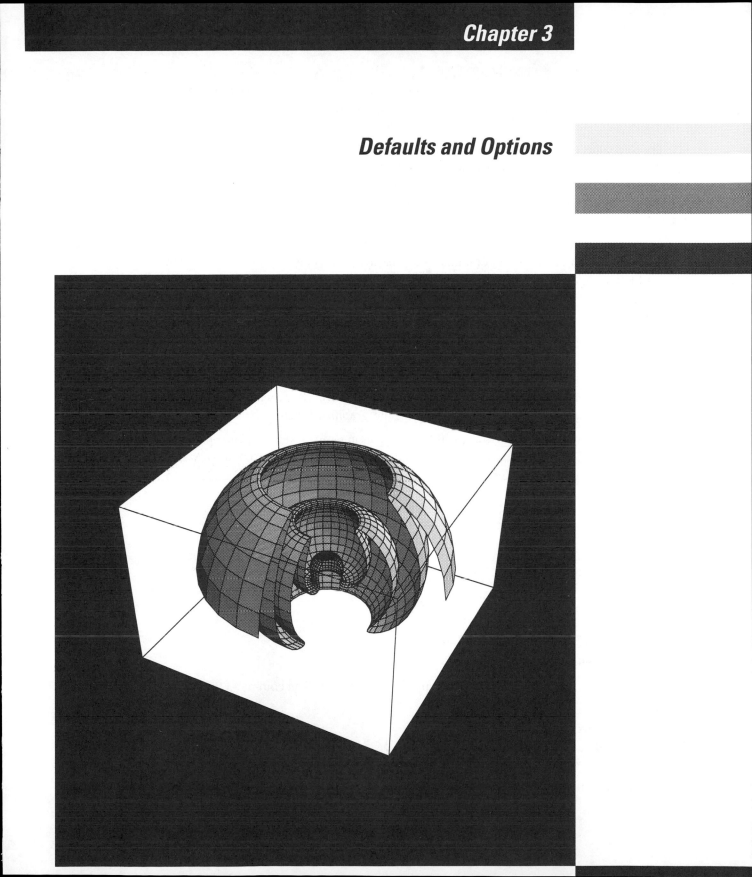

In a programming environment which is mainly used interactively the design of the user interface of functions that you write is of great importance. *Mathematica* provides two mechanisms for designing an easy to use interface.

The first one allows you to leave out parameters that you do not want to specify in simpler applications. Many built-in functions have default values for some of their arguments. These defaults are chosen to give the most basic form of the function in question. Novice users will not even have to know about these arguments; the function does the "right thing."

The second feature—options—is used if a function has many parameters that should be user-settable. In such a case, defaults would be too confusing. Options have a name and so their use is rather intuitive, while defaults rely on separate documentation about their placements and values. You can look at the default values of options and reset them globally.

Section 1 presents the basics about defaults for arguments. It also makes clear their limitations which motivate us to find ways to overcome them. These are discussed in section 2.

In section 3 we discuss how to define and use options for your own functions in the same way as for built-in functions, for example, the graphics functions.

The last section presents a useful utility for picking out options in an option list when we want to pass those options on to functions called within our function. A short discussion of sequences concludes this section.

About the illustration overleaf:

A cut-out view of a rotationally symmetric parametric surface. The command to generate this picture is

```
ParametricPlot3D[
    {r Cos[Cos[r]] Cos[psi], r Cos[Cos[r]] Sin[psi], r Sin[Cos[r]]},
    {r, 0.001, 9Pi/2 + 0.001, Pi/16}, {psi, 0, 3Pi/2, Pi/16}].
```

The equation of the generating curve in the x-z plane is $\phi = \cos r$.

■ 3.1 Simple Default Values for Arguments of Functions

■ 3.1.1 The Syntax of Defaults

To give a default value to a "blank" in a pattern, you simply use $x\_:v$ where v is the value that *Mathematica* should assume for the blank, named x, if it is left out from an expression. There is, however, another use of the colon in patterns; it is of the form $x:pattern$ and is used to give a name to a complicated pattern. Subsection 2.3.8 of the *Mathematica* book discusses the full syntax of patterns. You can experiment with *Mathematica* to understand the different interpretations quite easily using the function FullForm[] to see the internal form of an expression.

This is an optional pattern named x with default v.	In[1]:= **FullForm[x_:v]** Out[1]//FullForm= Optional[Pattern[x, Blank[]], v]
This is a pattern matching a list of two elements, and given the name t.	In[2]:= **FullForm[t:{_, _}]** Out[2]//FullForm= Pattern[t, List[Blank[], Blank[]]]
In the simplest case of giving a name to a blank, the colon is optional.	In[3]:= **FullForm[t:_]** Out[3]//FullForm= Pattern[t, Blank[]]
This gives the same expression.	In[4]:= **FullForm[t_]** Out[4]//FullForm= Pattern[t, Blank[]]

Defaults can therefore be given only to simple blanks. Here are the two forms that are possible.

$x\_:v$	an expression named x with default value v
$x\_h:v$	an expression with head h, named x and default value v

The two forms of defaults in patterns

■ 3.1.2 Possible Values for Defaults

The default value you specify is evaluated at the time the pattern is defined (in the left side of a rule normally). It cannot contain names of other patterns. This is a consequence of the way the right side of a rule is used. The pattern names are replaced by their values in parallel.

We try to define a function with a default value for its second argument. That value should be one less than the first argument. The function does nothing but return that value so we can check whether it works.	`In[5]:= f[n_, m_:(n 1)] := m`
However, this does not work.	`In[6]:= f[2]` `Out[6]= -1 + n`
It would be possible to make it work with this obscure construction. Using global variables in this way is, however, considered bad programming style.	`In[1]:= f[n_, m_:x] := Block[ {x = n-1}, m ]`
The default value of m is x, but inside the block x has a value and this value is then used.	`In[2]:= f[2]` `Out[2]= 1`

This kind of default is useful for constant default values, but not for more complicated cases. For defaults that depend on other parameters of a function, we need a different setup, which is the topic of the next section.

■ 3.2 Computed Defaults

■ 3.2.1 Using a Token

We have seen the preferred way of specifying computed defaults in section 1.5. The default value in the pattern is a special symbol whose presence is then checked inside the body of the rule in an If[] statement.

```
f[a_, b_, cv_:Automatic] :=
    Block[ {c = cv},
        If[ cv === Automatic, c = ... ];  (* compute default *)
        ⋮
    ]
```

Using computed defaults

The use of an extra local variable (c in this case) is necessary since names of patterns (cv in this case) cannot be used like local variables in the body of the rule. The symbol Automatic serves as a token to detect the use of the default value. It is important that it does not have a value. It is a built-in symbol and therefore protected, so everything is fine.

■ 3.2.2 Giving Several Rules

Another way of defining defaults was discussed in section 1.4. We can define a separate rule for each case we want to consider. One rule is the main one containing all the code that implements our function. It will require all arguments to be present. Other rules will have an argument list leaving some of the arguments of the main rule out. Their body is usually short. They simply compute the appropriate value for the left out arguments and call themselves again. Here is the general layout:

```
f[a_, b_, c_] :=
    Block[ {...},
        ⋮           (* main code goes here *)
        ]
f[a_, b_] :=
    Block[ {c},
        c = ... ;   (* compute value for c *)
        f[a, b, c]; (* call main routine   *)
    ]
```

Separate Rules for Default Arguments

As an example, let us add a new rule to our package **ComplexMap.m** from Chapter 1. When using the function `PolarMap[]`, the range for the angular variable will frequently be `{0, 2Pi}`, going once around the circle. We want to use this range by default. We add a rule for `PolarMap[]` that leaves out the second range specifier completely. For reference, we also include the outline of the previous rule for `PolarMap[]` and the new usage message.

```
PolarMap::usage = "PolarMap[f, {r0:0, r1, (dr)}, {phi0, phi1, (dphi)}, options...] plots
    the image of the polar coordinate lines under the function f. The default values
    of dr and dphi are chosen so that the number of lines is equal to the value of
    the option PlotPoints of Plot3D[]. The default for the phi range is {0, 2Pi}."

PolarMap[f_, {r0_:0, r1_, dr_:Automatic}, {phi0_, phi1_, dphi_:Automatic}, opts___Rule]:=
    Block[ {...},
        ⋮
    ]

PolarMap[ f_, rRange_List, opts___Rule ] := PolarMap[ f, rRange, {0, 2Pi}, opts ]
```

Part of **ComplexMap7.m**: A second rule for `PolarMap[]`

Observe that we can use a simpler form to match the radial range `rRange` in the second rule since all we want to do with it is to pass it along.

In[1]:= << ComplexMap7.m

This is the minimum information we have to specify, using all the defaults, including the start value 0 for the radius.
The function used in this example is called a *Möbius transform*.

In[2]:= PolarMap[(# - I)/(2# - 2)&, {4}]

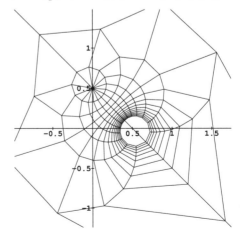

Increasing the number of radial lines to draw gives a much improved picture.

```
In[3]:= PolarMap[ (# - I)/(2# - 2)&,
              {4}, {-Pi, Pi, 2Pi/25} ]
```

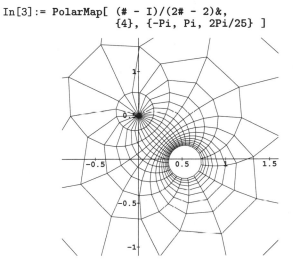

When you give several rules for the same function, you should think about the correct ordering of them. With complicated argument lists, *Mathematica* cannot always find out which of the rules is more special than the other and it might fail to reorder them accordingly. To verify that the ordering is correct, you should convince yourself that the argument list of a particular rule is not already matched by an earlier one. If it were, then these two rules should be exchanged.

■ 3.3 Defining Options for Functions

■ 3.3.1 Commands for Manipulating Options

Let us first review how you use options for built-in functions. The names and the default values of options are predefined. The command Options[*function*] returns a list of all options defined for the symbol *function* together with their default values. The result is returned as a list of rules and this form will be very useful when using options for our own functions. The command SetOptions[] is used to change these defaults.

Options[*f*]	list all options defined for *f*
SetOptions[*f*, *opt₁* -> *val₁*, ...]	change default values of options for *f*

The two functions for dealing with options

In order to define options for our own function *g*, we simply make an assignment for the expression Options[*g*]. This assignment is automatically stored with the symbol *g* rather than with Options. Once this is done we can use the function SetOptions[] in the same way as with built-in functions. This is briefly mentioned in subsection 2.3.7 of the *Mathematica* book.

This defines the names and initial default values of the options for g. Usually you should use = for this assignment rather than :=.

```
In[1]:= Options[g] = {Opt1 -> val1, Opt2 -> val2}
Out[1]= {Opt1 -> val1, Opt2 -> val2}
```

This works just as it does for built-in functions.

```
In[2]:= Options[g]
Out[2]= {Opt1 -> val1, Opt2 -> val2}
```

Mathematica tells us what it knows about g. The assignment from input line 1 has indeed been stored under the symbol g.

```
In[3]:= ?g
g
g/: Options[g] = {Opt1 -> val1, Opt2 -> val2}
```

This resets the default for the second option.

```
In[3]:= SetOptions[g, Opt2 -> newval]
Out[3]= {Opt1 -> val1, Opt2 -> newval}
```

Full error checking is done. Only previously defined options can be changed.

```
In[4]:= SetOptions[g, Opt3 -> val3]
SetOptions::missing: Opt3 is not an option for g.
Out[4]= SetOptions[g, Opt3 -> val3]
```

■ 3.3.2 Using Options in a Function

We now have a framework for dealing with options from the user's point of view. Inside our function, we need to look at the values that have been specified for each option, either the default value (if the option is not given on the parameter list) or the value given by an argument of the form *opt* -> *value*. Since options are given as rules, we use the standard substitution operation *expr* /. *rule* for extracting the value of an option. On the argument list, the options are specified by the pattern opt___Rule to match any sequence of options, including the empty sequence. Here is the outline of code to find the values of the options for a function:

```
BeginPackage["OptionUse`"]

g::usage = "g[n, options...] serves as an example for using options."

Opt1::usage = "Opt1 is an option of g[]."

Opt2::usage = "Opt2 is another option of g[]."

Options[g] = {Opt1 -> val1, Opt2 -> val2}

Begin["`Private`"]

g[ n_, opts___Rule ] :=
    Block[ {opt1, opt2},
        opt1 = Opt1 /. {opts} /. Options[g];
        opt2 = Opt2 /. {opts} /. Options[g];
        {n, opt1, opt2}
    ]

End[]

EndPackage[]
```

OptionUse.m: The use of options in a package

Let us understand how the the local variables opt1 and opt2 get their values. If g is called in the form g[5] without specifying any options, then the list {opts} is the empty list and the expression Opt1 /. {opts} evaluates to Opt1 since no rules were given on the right side of the substitution operator. The expression Options[g], however, is always a list of rules, one for each option, as we have seen. So the result of Opt1 /. {opts} /. Options[g] is the current default value for the option Opt1 which is then assigned to the local variable opt1. Note that substitutions are grouped to the left.

If, however, g is called in the form g[5, Opt1 -> newval] then the list {opts} is equal to {Opt1 -> newval} and the substitution Opt1 /. {opts} evaluates to newval. The following substitution newval /. Options[g] does nothing, since none of the rules matches. So the result of `Opt1 /. {opts} /. Options[g]` is the value for the option `Opt1` specified in the argument list which is then assigned to the local variable `opt1`.

Options should be documented just like functions. We have added such documentation to the example and also included the package framework. The names of the options are symbols in the package context. Our example function does nothing exciting. It simply returns the values of its required argument and the values of the options that are actually used inside the function. It is useful for playing around with options to understand how they work.

<table>
<tr><td></td><td><code>In[1]:= << OptionUse.m</code></td></tr>
<tr><td>Both options take on their default value.</td><td><code>In[2]:= g[5]</code>
<code>Out[2]= {5, val1, val2}</code></td></tr>
<tr><td>We can specify a different value for one of the options.</td><td><code>In[3]:= g[5, Opt1 -> 17]</code>
<code>Out[3]= {5, 17, val2}</code></td></tr>
<tr><td>We can also set a new default, as seen in subsection 3.3.1.</td><td><code>In[4]:= SetOptions[g, Opt2 -> newdef]</code>
<code>Out[4]= {Opt1 -> val1, Opt2 -> newdef}</code></td></tr>
<tr><td>This new value is now used instead of the old one.</td><td><code>In[5]:= g[5]</code>
<code>Out[5]= {5, val1, newdef}</code></td></tr>
<tr><td>The first value encountered on the argument list is used.</td><td><code>In[6]:= g[5, Opt2 -> value1, Opt2 -> value2]</code>
<code>Out[6]= {5, val1, value1}</code></td></tr>
</table>

■ 3.4 Filtering Options

In section 1.6 we have already seen an example of using options. The functions `CartesianMap[]` and `PolarMap[]` allowed optional arguments to be given. These options were not used inside the body of the function, but simply passed on to the `Show[]` command. There was one option of the function `Plot3D[]`, however, which was of interest to us. We used the value of the option `PlotPoints` to compute the default increment in the ranges. We cannot, however, override this default by including `PlotPoints->`*newval* as one of the arguments of `CartesianMap[]`. This fails for two reasons. First, the values of the options given as arguments are never looked at. This is easy to fix. In the preceding section, we have seen how to find the values of these options. In `CartesianMap[]` we would simply replace the line

```
plotpoints = PlotPoints /. Options[Plot3D]
```

by

```
plotpoints = PlotPoints /. {opts} /. Options[Plot3D].
```

The second problem is more subtle. `PlotPoints` is not a valid option for two-dimensional graphics (the `Graphics[]` command) and thus `Show[]` would give an error message if `PlotPoints` were part of the option sequence `opts` that is passed to `Show[]`.

This problem will occur in other places as well. Let us develop an auxiliary function for dealing with it in full generality.

■ 3.4.1 The Function FilterOptions

We need a function that takes as arguments the name of a command and a sequence of options and returns only those of the given options that are valid for this command, discarding all others.

`FilterOptions[`*f, options*`...]`	return only those options that are valid for the command *f*

A function for filtering option sequences

The functional programming style allows us to write this function in a very compact way.

```
BeginPackage["FilterOptions`"]

FilterOptions::usage = "FilterOptions[symbol, options...] returns a sequence
    of those options that are valid options for symbol."

Begin["`Private`"]

FilterOptions[ command_Symbol, opts___ ] :=
    Block[{keywords = First /@ Options[command]},
        Sequence @@ Select[ {opts}, MemberQ[keywords, First[#]]& ]
    ]

End[]
EndPackage[]
```

FilterOptions.m

To show you how it works, we use the same technique as in subsection 1.7.2. We assign sample parameters to the names of the patterns in the argument list for `FilterOptions[]` and then step through the body of the function, unwinding the nested function calls along the way. The symbol `Sequence` is perhaps new to you. The *Mathematica* book does not mention it by name. Sequences are indeed very peculiar objects, and a discussion of them is deferred to the next subsection. Suffice it to say that the expression substituted for a pattern of the form *name___* has as its head the symbol `Sequence`. Let us now understand what goes on when we evaluate the expression

  `FilterOptions[Graphics, Axes->None, PlotPoints->33, Framed->True]`.

We assign the value of the first parameter to its name.

```
In[1]:= command = Graphics

Out[1]= Graphics
```

And likewise for the sequence of options. Here we need the symbol `Sequence`.

```
In[2]:= opts = Sequence[Axes -> None, PlotPoints -> 33,
            Framed -> True]

Out[2]= (Axes -> None,PlotPoints -> 33,Framed -> True)
```

Here we unwind the initialization of the local variable `keywords`. First we get the list of options for our command. Graphics functions tend to have many options, and our list gets rather long.

```
In[3]:= Options[command]

Out[3]= {PlotRange -> Automatic,

                              1
    AspectRatio -> ─────────────,
                         GoldenRatio

    DisplayFunction :> $DisplayFunction,

    PlotColor -> Automatic, Axes -> None,

    PlotLabel -> None, AxesLabel -> None,

    Ticks -> Automatic, Framed -> False, Prolog -> {},

    AxesStyle -> {}, Epilog -> {}, Background -> Automatic,

    DefaultColor -> Automatic}
```

Then we extract the first element of each of the options. This gives a list of the *names* of the options.

```
In[4]:= keywords = First /@ %
Out[4]= {PlotRange, AspectRatio, DisplayFunction,
    PlotColor, Axes, PlotLabel, AxesLabel, Ticks, Framed,
    Prolog, AxesStyle, Epilog, Background, DefaultColor}
```

This is the result of mapping the predicate that occurs as the second argument of `Select[]` to our sequence of options. The predicate checks whether the names of these options occur in the list `keywords`. The second one of them is not a valid option for `Graphics[]`. The other two are.

```
In[5]:= MemberQ[keywords, First[#]]& /@ {opts}
Out[5]= {True, False, True}
```

This performs the selection according to the values of the predicate from the line above. Only the first and third of the options are selected.

```
In[6]:= Select[{opts}, MemberQ[keywords, First[#]]&]
Out[6]= {Axes -> None, Framed -> True}
```

This replaces the head `List` by `Sequence` making the result suitable for splicing into the options part of another function.

```
In[7]:= Sequence @@ %
Out[7]= (Axes -> None,Framed -> True)
```

It is easy to incorporate `FilterOptions[]` into the package **ComplexMap.m** from Chapter 1. We use hidden import (see subsection 2.2.2) since `FilterOptions[]` is of no use by itself and could be read in more than once without problem. The graphics function for which the options of `CartesianMap[]` and `PolarMap[]` are to be filtered is `Graphics`. Here is part of the new version, **ComplexMap8.m**:

```
BeginPackage["ComplexMap`"]

Needs["FilterOptions`"]
    ⋮
Begin["`Private`"]
    ⋮
CartesianMap[f_, {x0_, x1_, dx_:Automatic}, {y0_, y1_, dy_:Automatic}, opts___Rule] :=
    Block[ {x, y, points, plotpoints, ndx=N[dx], ndy=N[dy]},
        plotpoints = PlotPoints /. {opts} /. Options[Plot3D];
            ⋮
        Show[ MakeLines[points], FilterOptions[Graphics, opts],
            AspectRatio->Automatic, Axes->Automatic ]
    ]
PolarMap[f_, {r0_:0, r1_, dr_:Automatic}, {phi0_, phi1_, dphi_:Automatic}, opts___Rule]:=
    Block[ {r, phi, points, plotpoints, ndr=dr, ndphi=dphi},
        plotpoints = PlotPoints /. {opts} /. Options[Plot3D];
```

```
       ⋮
      Show[ MakeLines[points], FilterOptions[Graphics, opts],
          AspectRatio->Automatic, Axes->Automatic ]
   ]
     ⋮
End[(* "`Private`" *)]
EndPackage[]
```

Part of **ComplexMap8.m**

In[8]:= << ComplexMap8.m

The options `Axes` and `Framed` are passed to `Show[]`. The setting for `PlotPoints` takes effect immediately and is not passed on.

In[9]:= PolarMap[Exp, {1}, {-Pi, Pi}, PlotPoints->25,
            Axes->None, Framed->True]

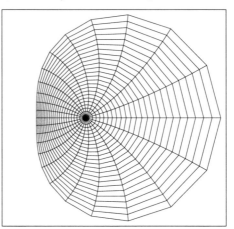

■ 3.4.2 Sequences

When an expression of the form $f[arg_1, arg_2, ...]$ is evaluated, the value of each of the arg_i becomes one element of the resulting expression. If, however, the value matched by a pattern of the form *name__* or *name___* is used in the position of one of the arg_i, then it is spliced in, occupying as many positions as there were elements in the matched expression. This is discussed in subsection B.5.1 of the *Mathematica* book. This object that causes its elements to be spliced in must, of course, be an expression itself, since all internal objects are expressions. The head of it is the symbol Sequence.

An example: The expression f[a, b, c, d] matches the pattern f[x_, opts___] with x becoming a as usual and opts becoming Sequence[b, c, d]. If opts is then used as an element of another expression, the sequence goes away and its elements are spliced in. So the expression g[u, opts, v] becomes g[u, b, c, d, v]. In the special case f[a] in which the matched sequence is empty, opts just becomes Sequence[]. If used as an element, this element simply goes away: g[u, opts, v] becomes g[u, v]!

An expression with head Sequence is very elusive. You cannot even look at it with FullForm[].

Here is an empty sequence.

```
In[1]:= Sequence[]
Out[1]= ()
```

The argument of FullForm[] goes away, giving a syntax error.

```
In[2]:= FullForm[%]
FullForm::argct: FullForm called with 0 arguments.
Out[2]= FullForm[]
```

About the only way to create a sequence of certain elements is to first create a list of these elements and then replace the head List by Sequence using the function Apply[*newhead*, *expression*]. In most cases, we prefer to write it in infix form as *newhead* @@ *expression*. We used this trick in the function FilterOptions[] in the preceding subsection.

Splicing in of sequences happens before any other evaluation, so you cannot give rules to change this behavior.

Chapter 4

Functional and Procedural Programming

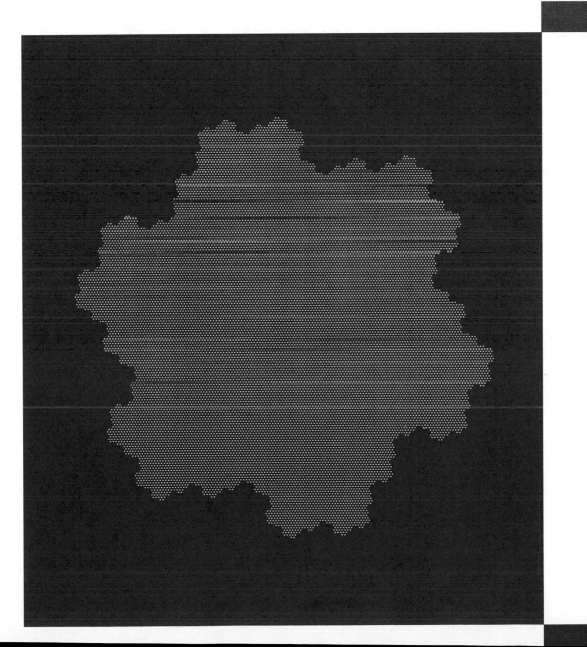

Mathematica allows you to program in a variety of styles. The most commonly identified programming styles are procedural programming, functional programming, declarative programming, and object oriented programming. If you have done your programming so far in one programming language exclusively, I could probably tell you which one you are used to by looking at your first *Mathematica* programs.

Rather than just continue in the style you are used to, I would like to show you how to use *Mathematica*'s commands in the best possible way, choosing the style that is best suited to the problem to be solved.

Chapter 4 of the *Mathematica* book gives you an introduction into some of these programming styles. Here, we put the emphasis on procedural and functional programming. In Chapter 6 we will look in detail into the unique style called *mathematical programming* that *Mathematica* offers.

Section 1 deals with the basic forms of iteration, available in almost any programming language. The next section then introduces some iteration constructs that are unique to *Mathematica*. They correspond very closely to the way we think about mathematics and should be used whenever possible.

Often we can avoid programming a loop altogether by applying functions to lists or other expressions. Section 3 and the following section about mapping of functions over expressions go to the heart of *Mathematica*'s programming language. Understanding this material will allow you to write concise programs in the style that the authors of *Mathematica* think is best to use.

In section 5 we will apply some of these ideas to an example. We will write functions to display and manipulate three-dimensional regular polyhedra.

Section 6 looks at some useful functions that deal with nested lists or matrices.

About the illustration overleaf:

This picture shows one step in an infinite iteration to generate a fractal tile—a shape that can be used to cover the whole plane without overlapping. It has the additional property that it can be cut into seven smaller copies of itself.

```
om7 = N[-1+Sqrt[-3]]/2; 17=om7-2
r7 = {0, 1,-1,om7,-om7,om7+1,-om7-1}
g7[x_] := Flatten[Outer[Plus, r7 , 17 x]]
points = Point[{Re[#],Im[#]}]& /@ Nest[g7, {0.}, 5];
graph4 = Graphics[Prepend[points, PointSize[0.003]]]
Show[graph4, AspectRatio->1, Axes->None, Framed->True].
```

■ 4.1 Loops

Loops and iterations are fundamental to any programming language. There are two basic kinds of iterations: repeating a statement a fixed number of times and iterating statements as long as a some condition is satisfied. This section introduces the commands `Do[]`, `While[]` and `For[]` that are available in one way or another in all procedural programming languages. As we will see later, *Mathematica* offers alternatives that are better suited for many applications.

■ 4.1.1 Iteration

The most basic form of a loop is `Do[`*statement, iterator*`]`. It allows you to perform a statement over and over again with the iterator variable taking on successive values. The `Do[]` statement itself returns no value.

This prints the squares of the first 5 positive integers. The output is a side-effect of the `Print[]` statement. No value is returned; there is no `Out[]` line.

```
In[1]:= Do[ Print[i^2], {i, 1, 5} ]
1
4
9
16
25
```

Sometimes you do not need the loop variable at all. The following piece of code computes the nth Fibonacci number f_n. Each Fibonacci number is defined as the sum of the previous two. The first and second one are defined to be one.

```
Fibonacci[1] = 1
Fibonacci[2] = 1

Fibonacci[n_Integer?Positive] :=
    Block[{fn1=1, fn2=1},
        Do[ {fn1, fn2} = {fn1 + fn2, fn1}, {n-2} ];
        fn1
    ]
```

Fibonacci1.m: Iterative computation of Fibonacci numbers

The local variables `fn1` and `fn2` are initialized to the first two Fibonacci numbers. In the loop we repeatedly replace `fn1` with the sum of the previous two numbers that are stored in `fn1` and `fn2` at any time. `fn2` is then set to the old value of `fn1`. The parallel assignment of both `fn1` and `fn2` makes this very easy without using an additional local variable. During the first iteration, we compute f_3 and so the loop has to be repeated $n - 2$ times.

This gives the Fibonacci number f_{100}. Loops like this are quite fast.

```
In[2]:= Fibonacci[100]
Out[2]= 354224848179261915075
```

Note that the number of iterations to be performed is computed when the iterator is evaluated before the loop is started. Changing the value of the upper bound in the iterator from inside the loop will therefore not change the number of iterations to be performed. The start, stop, and increment values can be general expressions. The only condition is that the number of iterations to perform evaluates to a number (other than a complex number). The increment need not divide evenly into the interval from start to stop. The number of iterations in the loop Do[*body*, {*start*, *stop*, *incr*}] is $(stop - start)/incr + 1$, rounded down to the nearest integer.

The number of steps is $(4a - 2a)/a + 1 \longrightarrow 3$, an integer.	`In[3]:= Do[ Print[i], {i, 2a, 4a, a} ]`
	
	`2 a`
	`3 a`
	`4 a`

The number of steps, $(3.5 - 0.0)/1 + 1 \longrightarrow 4.5$ is rounded to 4.	`In[5]:= Do[ Print[r], {r, 0.0, 3.5} ]`
	
	`0.`
	`1.`
	`2.`
	`3.`

In a nested loop, the iterator for the inner variable j is evaluated for each value of the outer variable i.	`In[7]:= Do[ Print[{i, j}], {i, 3}, {j, i} ]`
	
	`{1, 1}`
	`{2, 1}`
	`{2, 2}`
	`{3, 1}`
	`{3, 2}`
	`{3, 3}`

We have already encountered an example for the use of the Do[] in the function SplitLines[] in the package **ComplexMap.m** in subsection 1.7.2.

Do[] in *Mathematica* is similar to the DO loop in FORTRAN or BASIC and to the for loop in Pascal.

■ 4.1.2 Conditional Repetition of Statements

Often we want to perform a calculation repeatedly while a certain condition is true and stop as soon it becomes false. In this case, we do not know the number of iterations that are to be performed in advance and cannot use the Do[] loop, but use the While[] loop instead. Its form is While[*condition*, *body*]. Before each iteration the *condition* is tested. If the test returns True, the body of the loop is evaluated one more time, otherwise the loop terminates without returning a value. If the test does not return True the first time it is tested, the loop body is not executed at all.

For an example we look at the function $\pi(x)$ that, given a number x as argument, finds the number of primes less than x. Prime numbers are positive integers that have no divisors except 1 and themselves. The sequence starts with 2, 3, 5, 7, 11, ... The function `Prime[n]` returns the nth prime number p_n. We use iteration to find an n such that $p_n \le x < p_{n+1}$.

```
Attributes[PrimePi] = {Listable}

PrimePi[x_] := 0 /; x < 2

PrimePi[x_] :=
    Block[{li, n0, n1, m, nx = N[x]},
        li = LogIntegral[nx];
        n0 = Floor[li - LogIntegral[Sqrt[nx]]];
        n1 = Ceiling[li];
        While[ n1-n0 > 1,
            m = Floor[(n0+n1)/2];                      (* midpoint *)
            If[ Prime[m] <= nx, n0 = m, n1 = m ]
        ];
        n0
    ] /; x >= 2
```

PrimePi.m: Find the index of a prime number

The function `PrimePi[]` first gets an initial guess for n. It is a famous theorem in mathematics that the logarithmic integral gives a rather good guess for the n. In the range in which `Prime[]` is implemented this guess is always too large. On the other hand, subtracting the logarithmic integral of $\sqrt{x}$ gives an estimate that is too low. We therefore maintain two variables *n0* and *n1* that bracket the correct value. At the beginning, we know that the n we are looking for lies between *n0* and *n1*. At each iteration, we check whether the midpoint between them is too high or too low. If it is too low, we set *n0* to this midpoint otherwise we set *n1* to it. In each iteration, the interval between *n0* and *n1* is cut in half and after a few steps the two will differ by at most 1 and we have found n. Our algorithm does not work for arguments smaller than 2. We give a separate rule for this case. There are no primes smaller than 2 and so the value of `PrimePi[]` is 0 in this case.

The number 1993 is the 301st prime.	`In[2]:= PrimePi[1993]` `Out[2]= 301`
This computes the 100,000[th] prime.	`In[3]:= Prime[100000]` `Out[3]= 1299709`
And so we can test our function.	`In[4]:= PrimePi[%]` `Out[4]= 100000`
`PrimePi[]` is listable and it can therefore compute $\pi(x)$ for a whole list of numbers in one call.	`In[5]:= PrimePi[Range[10]]` `Out[5]= {0, 1, 2, 2, 3, 3, 4, 4, 4, 4}`

PrimePi[] is defined for all real numbers. We can therefore even plot it. The function jumps by one at prime values and is constant in between.

In[6]:= Plot[PrimePi[x], {x, 0, 20}]

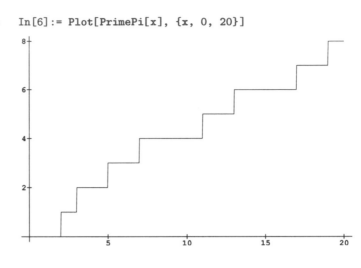

While[] in *Mathematica* is similar to the while loop in Pascal and C.

■ 4.1.3 The For Loop

The For[] loop is patterned after the corresponding loop in the language C. It does not have an equivalent in other languages. C lacks the equivalent of the Do[] loop and so expert C programmers often end up using For[] instead of Do[] to iterate over the values of a variable in *Mathematica*. Instead of Do[*body*, {*var*, *start*, *stop*}], they write For[*var=start*, *var*<=*stop*, *var*++, *body*], which looks a bit clumsy. The For[] loop is useful in more complicated instances where the iteration involves several variables and end conditions. The For[] loop can easily be expressed in terms of a While[] loop. Instead of

$$\text{For}[start,\ test,\ step,\ body]$$

we can write

$$start;\ \text{While}[test,\ body;\ step].$$

If you are not familiar with For[], this correspondence can help you understand how it works.

■ 4.1.4 Testing the Exit Condition at the End of a Loop

Besides the While[] loop that checks a condition at the beginning of each iteration, many programming languages also offer a loop that checks the condition *after* each iteration. Here are several ways in which we could implement a hypothetical Until[*body*, *test*] that repeats *body* until *test* becomes True.

```
While[ True, body; If[test, Break[]] ]
                              break from an infinite While[] as soon as test is true.
t=False; While[ !t, body; t=test ]
                              remember the value of test for the next iteration.
For[ t=False, !t, t=test, body ]
                              use a For[] loop.
```

Loops that check their test at the end of an iteration

If we wanted, we could define our own command Until[] in terms of one of the possibilities outlined.

```
Until::usage = "Until[body, test] evaluates body until test becomes true."

Begin["`Private`"]

Attributes[Until] = {HoldAll}

Until[body_, test_] := Block[ {`t}, For[ t=False, !t, t=test, body ] ]

End[]
Null
```

Until.m: A loop that checks its test after each iteration

The attribute HoldAll prevents evaluation of the parameters of Until[]. They are only evaluated inside For[]. All loops should behave like this. The definition of Until[] is so simple that it is not even worth creating a separate package context for it. We do, however, isolate the local variable t in a private subcontext of the current context. Note that in this case (in which we do not set up a package context) the first occurrence of the local variable t is prefixed by a context mark, like this: `t. This makes sure that the symbol t is created in the current context, even if it already exists in the global context.

This is a fixed point iteration. It sets x to $1 + 1/x$ until x is equal to $1 + 1/x$. This number is called the *golden ratio*. More about fixed points in subsection 4.3.3. Note that the precision of the result is greater than the precision of the starting point.

```
In[2]:= x = N[1, 20]; Until[ x=1+1/x, x==1+1/x ]; x

Out[2]= 1.6180339887498948482045868343656381177203091798057\
   6286213544862099
```

Loops that test their condition at the end are particularly useful for interactive input from the user. We will repeatedly prompt the user for input until he has given a valid answer. The following code will insist in the user typing a number that satisfies a given predicate (A predicate is a function that returns `True` or `False`).

```
GetNumber[prompt_String, predicate_:(True&)] :=
    Block[{answer},
        While[ True,
            answer = Input[prompt];
            If[ NumberQ[answer] && predicate[answer], Break[] ];  (* good    *)
            If[ answer === EndOfFile, Break[] ];                  (* escape *)
            Print["Please enter a number that satisfies ", predicate]
        ];
        answer
    ]
```

GetNumber.m: Repeatedly prompting the user for input

It is a good idea to allow some escape mechanism from the loop that prompts for input. In this case, generating an end-of-file character will exit the loop. If the answer was rejected, a friendly message is printed. The default value for the predicate is the pure function `True&` that returns `True` independent of its argument, and so it will accept any number.

An input that does not satisfy the predicate will issue a message and prompt again.

```
In[1]:= input = GetNumber["Enter a prime: ", PrimeQ]

Enter a prime: 9
Please enter a number that satisfies PrimeQ
Enter a prime: 11

Out[1]= 11
```

Mathematica can read any expression. You can perform arbitrary calculations in the input you type. If it does evaluate to a number, it will be accepted.

```
In[2]:= GetNumber["Enter any number: "]

Enter any number: junk
Please enter a number that satisfies True &
Enter any number: (-1 + 5 I)^2

Out[2]= -24 - 10 I
```

`Until[]` is similar to `repeat...until` in Pascal and to `do...while` in C (with the truth value of the test reversed!).

■ 4.2 Structured Iteration

In many programming languages the loops presented in section 4.1 are all that is available. The structured iteration commands of this section are perhaps new to you. If you have programmed in LISP or APL, then you will, however, recognize many familiar and useful commands. The flow-control statements of traditional languages were not designed with the applications in mind that are now possible in *Mathematica*, but rather they were selected for ease of implementation and the need of applications in computer science itself.

Since *Mathematica* does offer the traditional looping constructs, as we have just seen, it is rather tempting to simply continue using these familiar means of flow-control instead of using a more natural, problem-oriented approach. In this section, I would like to show you the transformation from the old approach to *Mathematica*'s way of functional programming.

■ 4.2.1 Sums and Products

Given the problem of adding the square roots of the first 500 integers, the solution in most programming languages is to use an auxiliary variable that is incremented by the square root of a loop index iterating from 1 to 500.

```
sum = 0.0;
Do[ sum = sum + N[Sqrt[i]], {i, 1, 500} ];
sum
```

The procedural way of adding numbers

In *Mathematica*, this loop reduces to a single statement that directly corresponds to the mathematical formula $\sum_{i=1}^{500} \sqrt{i}$.

```
Sum[ N[Sqrt[i]], {i, 1, 500} ]
```

The mathematical way of adding numbers

Mathematica does not force you to think about *how* to implement a summation, but lets you focus on the concept itself instead. Product[] works in the same way, multiplying its terms together instead of adding them up.

This computes the product $\prod_{i=0}^{5}(x-i)$ and expands it.

```
In[1]:= Product[ x-i, {i, 0, 5} ] // Expand
                    2        3        4        5    6
Out[1]= -120 x + 274 x  - 225 x  + 85 x  - 15 x  + x
```

The iterator used for sums and products is the same as that used for loops (see subsection 4.1.1). This allows a rather sophisticated way of computing the same product.

```
In[2]:= Product[ e, {e, x, x-5, -1} ] // Expand
                    2        3        4        5    6
Out[2]= -120 x + 274 x  - 225 x  + 85 x  - 15 x  + x
```

■ 4.2.2 Tables

`Sum[]` and `Product[]` are two examples of a class of commands that evaluate an expression several times varying one or several index variables and then collecting the results in a specific way, either adding them up or multiplying them together. The `Table[]` command is the simplest of them. It just collects its results in a list.

This generates a list of the first 10 powers of x.

```
In[1]:= Table[ x^i, {i, 0, 9} ]
                   2   3   4   5   6   7   8   9
Out[1]= {1, x, x , x , x , x , x , x , x , x }
```

This generates a list of the first 6 of the polynomials used in the previous example. Note how the inner index (in the product) depends on the outer index (of the table).

```
In[2]:= Table[ Product[x-j, {j, 0, i}], {i, 0, 5} ]
Out[2]= {x, (-1 + x) x, (-2 + x) (-1 + x) x,
   (-3 + x) (-2 + x) (-1 + x) x,
   (-4 + x) (-3 + x) (-2 + x) (-1 + x) x,
   (-5 + x) (-4 + x) (-3 + x) (-2 + x) (-1 + x) x}
```

`TableForm[]` prints the elements of a list on separate lines as a table.

```
In[3]:= TableForm[%]
Out[3]//TableForm=
   x
   (-1 + x) x
   (-2 + x) (-1 + x) x
   (-3 + x) (-2 + x) (-1 + x) x
   (-4 + x) (-3 + x) (-2 + x) (-1 + x) x
   (-5 + x) (-4 + x) (-3 + x) (-2 + x) (-1 + x) x
```

`Table[]`, like all of these structured iterators, can have more than one iterator specification, giving you multidimensional tables. Two iterators generate matrices. The first iterator is the outermost one. `Table[`*expr*`,` *iterator₁* `,` *iterator₂* `]` is therefore equivalent to `Table[Table[`*expr*`,` *iterator₂* `],` *iterator₁* `]`.

The matrix of the first n powers of n different variables is called *Vandermonde*'s matrix.

```
In[4]:= Table[ x[i]^j, {i, 1, 5}, {j, 0, 4} ] //
        MatrixForm
```

$$
\mathrm{Out[4]//MatrixForm=}
\begin{matrix}
1 & x[1] & x[1]^2 & x[1]^3 & x[1]^4 \\
1 & x[2] & x[2]^2 & x[2]^3 & x[2]^4 \\
1 & x[3] & x[3]^2 & x[3]^3 & x[3]^4 \\
1 & x[4] & x[4]^2 & x[4]^3 & x[4]^4 \\
1 & x[5] & x[5]^2 & x[5]^3 & x[5]^4
\end{matrix}
$$

Its determinant in expanded form has $n!$ terms. Here n is 5, and we get 120 terms—too much to print out in full.

```
In[5]:= Det[%] // Short
Out[5]//Short=
```
$$
x[1]^4\, x[2]^3\, x[3]^2\, x[4] + \langle\langle 118 \rangle\rangle + x[2]\, x[3]^2\, x[4]^3\, x[5]^4
$$

But it factors in this nice way.

```
In[6]:= Factor[%]
Out[6]= (x[1] - x[2]) (x[1] - x[3]) (x[2] - x[3])
    (x[1] - x[4]) (x[2] - x[4]) (x[3] - x[4]) (x[1] - x[5])
    (x[2] - x[5]) (x[3] - x[5]) (x[4] - x[5])
```

■ 4.2.3 Arrays

To understand the difference between `Array[]` and the other structured iterators we have discussed so far, let's have a closer look at how these iterators work. Assume the following iterator: `Table[`*expr*`, {i,` *start*`,` *final*`}]`, where the iterator variable is `i`. First the variable is set to the first value, *start*. This is done in much the same way as if you had typed `i=`*start*. Then *expr* is evaluated, using the current value for `i`, wherever `i` occurs in *expr*. For the next iteration, the next value *start*`+1` is assigned to `i` and *expr* is again evaluated.

Another way to look at this is to say that the expression *expr* describes a function of the iterator variable `i`. For each value of `i`, we get a value $f[\mathtt{i}]$ for some function f. The iterator `Array[]` takes this point of view. In `Array[`f`,` n`]`, f is the *name* of a function that is applied to each value of the iterator in turn. We do not need a named iterator variable as the name of the parameter of a function does not matter at all. *Mathematica* simply generates the expressions $f[1], f[2], \ldots, f[n]$ and evaluates them.

You could express this action using `Table[]`: The expression `Array[`f`,` n`]` is equivalent to `Table[`$f[\mathtt{i}]$`, {i,` n`}]` (Assuming that there is no conflict of variable names, i.e., `i` does not appear in the definition of f).

This generates a list of the first 10 prime numbers.

```
In[1]:= Array[ Prime, 10 ]
Out[1]= {2, 3, 5, 7, 11, 13, 17, 19, 23, 29}
```

This is the equivalent `Table[]` command.

```
In[2]:= Table[ Prime[i], {i, 10} ]
Out[2]= {2, 3, 5, 7, 11, 13, 17, 19, 23, 29}
```

You can also express a `Table[]` in terms of an `Array[]`. If the expression to be tabulated is not of the form $f[var]$, where *var* is the iterator variable, you need to express it as a pure function. For example, `Table[i^2, {i, 10}]` can be expressed as `Array[#^2&, 10]` (For more on pure functions, see section 5.2).

`Array[]` is not as flexible as `Table[]`. The increment of the iterator is always one. The iterator starts at 1 unless a different starting point is given as a third argument, as in `Array[f, 4, 0]` $\longrightarrow$ `{f[0], f[1], f[2], f[3]}`. A multidimensional array has the form `Array[`f`, {`n_1`, `n_2`,..., `n_k`}]` and corresponds to the table

$$\text{Table[} f[i_1, i_2, \ldots, i_n], \{i_1, n_1\}, \{i_2, n_2\}, \ldots, \{i_k, n_k\} \text{]}.$$

This array gives again Vandermonde's matrix from subsection 4.2.2. We have two iterators and so we need a pure function of two variables.

```
In[4]:= Array[ x[#1]^(#2-1)&, {5, 5} ] // MatrixForm
```

$$
\begin{matrix}
1 & x[1] & x[1]^2 & x[1]^3 & x[1]^4 \\
1 & x[2] & x[2]^2 & x[2]^3 & x[2]^4 \\
1 & x[3] & x[3]^2 & x[3]^3 & x[3]^4 \\
1 & x[4] & x[4]^2 & x[4]^3 & x[4]^4 \\
\end{matrix}
$$

Out[4]//MatrixForm= $\begin{matrix} 1 & x[5] & x[5]^2 & x[5]^3 & x[5]^4 \end{matrix}$

■ 4.3 Function Application Instead of Iteration

■ 4.3.1 Mapping a Function over a List

Quite often we want to apply some function to a list of expressions. Consider the example of squaring all the numbers in a list. A procedural program would iterate over the list and build up a new list of the results:

```
SquareList[l_List] :=
    Block[{result = {}, i},
        Do[ AppendTo[result, l[[i]]^2], {i, Length[l]} ];
        result
    ]
```

Using a loop to apply a function to the elements of a list

Building up the result by using Append[] or AppendTo[] when you know the length of the result beforehand is not a good solution. It is similar to using an auxiliary variable for adding up the the terms in a sum (see subsection 4.2.1). In the last section, we saw how to improve such iterations by using a structured iterator. Since we want a list of the results, we use Table[]:

```
SquareList[l_List] :=
    Block[{i},
        Table[ l[[i]]^2, {i, Length[l]} ]
    ]
```

Using a structured iterator to apply a function to the elements of a list

In each of these cases, we apply the same function to all the elements of the original list. In such cases we should use Map[*f*, *list*] to perform the operation. Map[] takes the name of a function *f* and applies it to all the elements of *list*. The function we want is "square the argument". There is no built-in function for this, so we use a pure function instead. The pure function to use in this case is #^2&.

```
SquareList[l_List] := Map[ #^2&, l ]
```

Mapping a function at the elements of a list

This operation is so common that most of the built-in functions map themselves automatically over lists. This property is called *listability*. Taking advantage of this, our example becomes even simpler.

```
SquareList[l_List] := l^2
```

Using the built-in listability of `Power[]`

■ 4.3.2 Listability

It is worth understanding exactly what happens if a listable function has more than one argument. The internal form of `l^2` is `Power[l, 2]`. Let us assume that the value of `l` is `{a, b, c}`, and so our expression is `Power[{a, b, c}, 2]`. The first argument of `Power[]` is a list, while its second argument is a number. In this case, the expression is transformed into a list of powers, duplicating the second argument, and we get `{Power[a, 2], Power[b, 2], Power[c, 2]}`, or `{a^2, b^2, c^2}`. If all the arguments of a listable function are lists, then their elements are picked in parallel. `Power[{a, b, c}, {1, 2, 3}]`, or `{a, b, c}^{1, 2, 3}` in the usual notation, becomes `{Power[a, 1], Power[b, 2], Power[c, 3]}`, or `{a, b^2, c^3}`. In this case, all the arguments must have the same length. Other examples of this kind are `{a, b, c} + 1` $\longrightarrow$ `{1 + a, 1 + b, 1 + c}` and `{1, 2, 3} {x, y, z}` $\longrightarrow$ `{x, 2 y, 3 z}`.

You can declare your own functions to be listable by giving them the attribute `Listable`. We have already given an example of this in the function `PrimePi` in subsection 4.1.2.

Some functions behave like listable ones but they do not have the attribute `Listable`. The reason is that sometimes a list as an argument has a special meaning. Consider differentiation. The general form is `D[`*expr*`, {`*var*`, `*n*`}]` to differentiate *expr* *n* times with respect to *var*. The list in the second parameter has a special meaning. If the first argument is a list, then we still want each of its elements to be differentiated *n* times with respect to *var*. If `D[]` had the attribute `Listable` then `D[{e1, e2}, {x, 3}]` would turn into `{D[e1, x], D[e2, 3]}` which does not make sense. What we want is `{D[e1, {x, 3}], D[e2, {x, 3}]}`.

You can give a simple rule for one of your own functions to make it behave in the same way.

```
f[ l_List, args___ ] := Map[ f[#, args]&, l ]
f[ e_,... ] := ... (* code for the usual case *)
```

A function that is listable over its first argument only

This rule works for functions of any number of additional arguments because we have used the *triple blank* pattern `arg___` to match any number of elements.

The following example defines rules for diff[*expr*, *var₁*, *var₂*, ..., *var_n*], a function that differentiates its first argument once with respect to each of the variables given as additional arguments.

Make diff listable over its first argument.

```
In[1]:= diff[l_List, args___] := Map[ diff[#, args]&, l ]
```

Differentiate once with respect to the first variable.

```
In[2]:= diff[e_, var_, rest___] := diff[ D[e, var], rest ]
```

No differentiation, if no variables are given.

```
In[3]:= diff[e_] := e
```

Differentiate the three functions x, x^y, and x y with respect to x and then with respect to y.

```
In[4]:= diff[{x, x^y, x y}, x, y]
                 -1 + y    -1 + y
Out[4]= {0, x         + x         y Log[x], 1}
```

The rules we have given are of course none other than the ones already built into D[].

```
In[5]:= D[{x, x^y, x y}, x, y]
                 -1 + y    -1 + y
Out[5]= {0, x         + x         y Log[x], 1}
```

■ 4.3.3 Iterated Application of a Function

Another common task is to repeatedly apply a function to an expression. An example is Newton's iteration formula for approximating the square root of a number r. Given an approximation x_i, we can find a better approximation by computing $x_{i+1} = (x_i + r/x_i)/2$. Starting with an initial guess x_0 we get a sequence of better and better approximations $x_1, x_2, \ldots$ that converges to $\sqrt{r}$.

We want to compute an approximation to $\sqrt{2}$ starting with the initial guess 1. We want the computations to be accurate to 40 digits.

```
In[1]:= r = 2; x = N[1, 40];
```

This computes and prints the first 7 iterations.

```
In[2]:= Do[ x = (x + r/x)/2; Print[x], {7}]
1.5
1.4166666666666666666666666666666666666667
1.4142156862745098039215686274509803921569
1.4142135623746899106262955788901349101166
1.4142135623730950488016896235025302430615
1.4142135623730950488016887242096980785967
1.4142135623730950488016887242096980785967
```

The last value of x is accurate within the precision of 40 digits.

```
In[3]:= x - N[Sqrt[r], 40]

Out[3]= 0.
```

The function Nest[*f*, *x*, *n*] allows us to apply a function *f* a given number times *n* to an initial value *x*. Using this, we do not need to program a loop as we just did. We

need to express the formula for the new x as a function of the old x. The pure function
`(# + r/#)/2&` does this.

Initialize r.	`In[4]:= r = 2;`

This performs 7 iterations and returns the last result. It is the same as in the previous calculation.	`In[5]:= Nest[ (# + r/#)/2&, N[1, 40], 7 ]` `Out[5]= 1.4142135623730950488016887242420969807856967`

How many iterations are necessary to guarantee accuracy to the number of digits that
we have specified? One way to find out when to stop the iteration is to compare the
result of the last iteration with the previous value of x and stop if that difference gets
smaller than the precision with which we do the calculations. *Mathematica* can detect
this condition by itself. The function `FixedPoint[`f`, `x`]` keeps applying f to x just
like `Nest[]`, but it stops as soon as the expression no longer changes. A value x such
that $f(x)$ is equal to x is called a *fixed point* of f.

Initialize r.	`In[6]:= r = 2;`

This performs as many iterations as are necessary to find the fixed point of `(# + r/#)/2&` and returns the last result.	`In[7]:= FixedPoint[ (# + r/#)/2&, N[1, 40] ]` `Out[7]= 1.4142135623730950488016887242420969807856967`

If we modify our function to print the value of its argument, we can find out the intermediate values and can count the number of iterations performed.	`In[8]:= FixedPoint[ (Print[#]; (# + r/#)/2)&, N[1, 40] ]`
	`1.`
	`1.5`
	`1.4166666666666666666666666666666666666667`
	`1.4142156862745098039215686274509803921569`
	`1.4142135623746899106262955788901349101166`
	`1.4142135623730950488016896235025302430243615`
	`1.4142135623730950488016887242420969807856967`
	
	`Out[8]= 1.4142135623730950488016887242420969807856967`

The cosine function also has a fixed point. This iteration is however much slower because the rate of convergence is smaller than in Newton's iteration.	`In[9]:= FixedPoint[ Cos, N[1/2, 40] ]` `Out[9]= 0.73908513321516064165531208767387340 40257`

The cosine of the last result is in fact equal to it.	`In[10]:= Cos[%] == %` `Out[10]= True`

You have to be careful when using `FixedPoint[]`. If a function does not have a
fixed point, or if the iteration does not converge, *Mathematica* gets trapped in an infinite
loop. If this happens, you can interrupt and abort the calculation. As a precaution,
you can give a third argument to `FixedPoint[]` that specifies the maximum number of
iterations to perform.

■ 4.3.4 Application: Newton's Iteration Formula in General

In the preceding subsection we have briefly alluded to Newton's iteration formula for finding square roots of functions. The method can be used more generally to find zeros of functions. A zero of a function is of course a number x such that $f(x) = 0$. The Newton iteration proceeds by finding better and better approximations to a zero of f. Given an approximation x_i we find a better one with the formula $x_{i+1} = x_i - f(x_i)/f'(x_i)$.

The package **Newton.m** defines two functions, NewtonZero[], to find zeros of functions, and NewtonFixedPoint[], to find fixed points of functions. NewtonZero[] implements Newton's iteration formula. It is called as NewtonZero[*expr, var, start*], in which you give an expression *expr* involving the variable *var* and a starting point *start* for the iteration. It can also be called as NewtonZero[*f, start*], in which you give the name *f* of a function and again a starting point *start* for the iteration. The command NewtonFixedPoint[*f, start*] finds a fixed point of the function *f* starting the iteration at *start*. Both functions perform up to $RecursionLimit many iterations to find a zero or fixed point of the given function. If they cannot find one within the precision of the starting point given, they print a message and return the value found so far.

```
BeginPackage["Newton`"]

NewtonZero::usage = "NewtonZero[expr, x, x0] finds a zero of expr as a function of x
    using the initial guess x0 to start the iteration. NewtonZero[f, x0] finds
    a zero of the function f. The recursion limit determines the
    maximum number of iteration steps that are performed."

NewtonFixedPoint::usage = "NewtonFixedPoint[f, x0] finds a fixed point
    of the function f using the initial guess x0 to start the iteration."

Newton::noconv = "Iteration did not converge in `1` steps."

Begin["`Private`"]

NewtonZero[expr_, x_, x0_] := NewtonZero[ Function[x, expr], x0 ]

NewtonZero[f_, x0_] :=
    Block[{res, x0a, prec = Precision[x0], fp = f'},
        x0a = SetPrecision[x0, prec + 10];
        res = FixedPoint[(# - f[#]/fp[#])&, x0a, $RecursionLimit];
        x0a = f[res];
        If [ Accuracy[x0a] - Precision[x0a] < prec,
            Message[Newton::noconv, $RecursionLimit] ];
        N[res, prec]
    ]

NewtonFixedPoint[f_, x0_] := NewtonZero[(f[#] - #)&, x0]

End[]

Protect[ NewtonZero, NewtonFixedPoint]
```

```
EndPackage[]
```

Newton.m: Newton's iteration formula

Let us look at the rule for `NewtonZero[f_, x0_]`. It first finds the precision of x0 and then artificially raises the precision of the starting point by some amount so that the fixed-point calculations are performed at a higher precision. It then computes the derivative `f'` of the given function `f`. We assign the result to a local variable so that it does not need to be recomputed at each iteration step. We then find the fixed point of the pure function `(# - f[#]/fp[#])&` which corresponds to Newton's formula $x_i - f(x_i)/f'(x_i)$. After returning from the fixed point calculation, we compute `f[res]` which will be close to zero if we did find the fixed point. The expression `Accuracy[x0a]-Precision[x0a]` finds the number of zeros to the right of the decimal point in x0a (more about this in subsection 7.1.2). If this number is less than the precision we started with, we print a message. Finally, we lower the precision of the result back to the original precision of the input.

If `NewtonZero[]` is called in the form `NewtonZero[`*expr, var, start*`]`, then the function whose zero we want to find is given as an expression *expr* that is to be considered a function of *var*. We can make it into the pure function `Function[`*var, expr*`]` and then simply use the other rule because we now have a named function.

To find a fixed point of the function $f(x)$, we find a zero of the function $f(x) - x$. `NewtonFixedPoint[`*f, start*`]` therefore simply calls `NewtonZero[]` with the pure function $(f[\#] - \#)\&$.

If you call `NewtonFixedPoint[`*f, start*`]` with a pure function itself, the program goes through many levels of constructing new pure function from old ones. Nevertheless, it is capable of differentiating them correctly.

This gives again the fixed point of the cosine function, but much faster than in subsection 4.3.3 on page 84.

```
In[2]:= NewtonFixedPoint[ Cos, N[1, 40] ]
Out[2]= 0.7390851332151606416553120876738734040134
```

Here is the golden ratio to 20 digits, much faster than in subsection 4.1.4 on page 74.

```
In[3]:= NewtonFixedPoint[ 1 + 1/#&, N[2, 20] ]
Out[3]= 1.6180339887498948482
```

The function $z^2 + 1$ does not have any real zeroes and so the iteration cannot converge.

```
In[4]:= NewtonZero[ z^2 + 1, z, 0.5 ]
Newton::noconv: Iteration did not converge in 256 steps.
Out[4]= -4.781570434570313
```

If we give a complex starting point, however, it converges to one of the two complex zeros.

```
In[5]:= NewtonZero[ z^2 + 1, z, 0.5 + I ]
Out[5]= 1. I
```

A starting point with a negative imaginary part converges to the other complex zero.

```
In[6]:= NewtonZero[ z^2 + 1, z, 0.5 - I ]
Out[6]= -1. I
```

The built-in functions `NRoots[]` and `FindRoot[]` use similar techniques to find numerical solutions to equations.

```
In[7]:= NRoots[ z^2 + 1 == 0, z ]
Out[7]= z == -1. I || z == 1. I
```

■ 4.3.5   Returning the Intermediate Results

In subsection 4.3.3, we have seen how to repeatedly apply a function to an argument. `Nest[]` returns the final value. We can get a list of all the intermediate values with `NestList[]`.

This shows the basic idea. Note that the result is a list of length 5 and not 4.

```
In[1]:= NestList[f, a, 4]
Out[1]= {a, f[a], f[f[a]], f[f[f[a]]], f[f[f[f[a]]]]}
```

This is an interesting way of generating a 5 by 5 identity matrix.

```
In[2]:= NestList[RotateRight, {1,0,0,0,0}, 4] //
           MatrixForm
                    1  0  0  0  0
                    0  1  0  0  0
                    0  0  1  0  0
                    0  0  0  1  0
Out[2]//MatrixForm= 0  0  0  0  1
```

Using `NestList[]` instead of `Nest[]` in input line 5 on page 84 shows us the intermediate results.

```
In[3]:= NestList[ (# + 2/#)/2&, N[1, 40], 7 ] //
           TableForm
Out[3]//TableForm=
 1.
 1.5
 1.41666666666666666666666666666666666667
 1.4142156862745098039215686274509803921569
 1.4142135623746899106262955788901349101166
 1.4142135623730950488016896235025302343615
 1.4142135623730950488016887242096980785696
 1.4142135623730950488016887242096980785696
```

■ 4.4 Map and Apply

Not all programming languages give you the ability to treat functions like any other objects (symbols or numbers). In *Mathematica*, you can assign them to variables and they can be arguments to other functions. Two important commands that take functions as arguments are `Map[]` and `Apply[]`. We have already used them in a few places, but now we want to look at them in detail.

■ 4.4.1 Mapping Functions onto Expressions

In subsection 4.3.1 we have briefly encountered `Map[]`. `Map[`*f*, *list*`]` maps the function *f* over the elements of the list *list*. That means that it forms the expression $f[e_i]$ for each element e_i of *list* and returns the list of the results. The second argument of `Map[]` need not be a list however. Any expression of the form *head*`[`e_1, e_2, ..., e_n`]` will do. The result of the mapping is the expression *head*`[`$f[e_1]$, $f[e_2]$, ..., $f[e_n]$`]`.

The internal form of the second argument is `Plus[a, b, c]` and so we get `Plus[f[a], f[b], f[c]]`, which is printed as shown.

```
In[1]:= Map[ f, a + b + c ]

Out[1]= f[a] + f[b] + f[c]
```

The pure function used here extracts the second element of its argument. The result of the mapping is to extract the second element from each element of the list.

```
In[2]:= Map[ #[[2]]&, {a+b, 2 x y, z^2,
                 alpha (gamma + delta) / beta} ]

                           1
Out[2]= {b, x, 2, ————}
                          beta
```

`FullForm[]` is useful to understand what the second element is in each of the four elements of our list. Sometimes it is not what it appears to be (look at the fourth element).

```
In[3]:= FullForm[ {a+b, 2 x y, z^2,
                 alpha (gamma + delta) / beta} ]

Out[3]//FullForm=
 List[Plus[a, b], Times[2, x, y], Power[z, 2],
   Times[alpha, Power[beta, -1], Plus[delta, gamma]]]
```

In the package **ComplexMap.m** in Chapter 1 we used `Map[]` with its optional third argument to specify a level other than 1 at which to map. In a matrix, the entries are at level 2 (there are two levels of lists) and if we want to map a function at these entries, we can use `Map[`*f*, *matrix*, `{2}]`. Note the difference between this and `Map[`*f*, *matrix*, `2]` which would map at all levels *up to* 2. Level specifications are explained in subsection 2.1.10 of the *Mathematica* book.

■ **4.4.2 Map at Particular Positions**

Map[] always maps the function at all elements of the given levels. The command MapAt[*f, expr, poslist*] allows you to map a function at any given positions in your expression. (Positions and parts of expressions are treated in more detail in subsection 2.1.11 of the *Mathematica* book.) *poslist* is a single position or a list of positions. The positions are lists of numbers that describe how to descend down the expression tree to that place.

This maps f at the first element of the second element (the symbol c).

```
In[4]:= MapAt[f, a b + c d + e f, {2,1}]

Out[4]= a b + e f + d f[c]
```

This maps the square root function at position {1} (the term a b), at position {2, 2} (the symbol c) and at position {3, 1} (the number 4). The usual evaluation rules then take effect to make the simplifications
Sqrt[a b] $\longrightarrow$ Sqrt[a] Sqrt[b] and
Sqrt[4] $\longrightarrow$ 2.

```
In[5]:= MapAt[ Sqrt, a b + 2 c + 4/x,
               {{1}, {2, 2}, {3, 1}}]
```

$$Out[5]= Sqrt[a]\ Sqrt[b] + 2\ Sqrt[c] + \frac{2}{x}$$

Here is an expression.

```
In[8]:= expr = a + b/a + c E^(a+1)
```

$$Out[8]= a + \frac{b}{a} + E^{1 + a}\ c$$

If we wanted to map a function f at all occurrences of a, we first find all the positions of a in our expression.

```
In[9]:= Position[expr, a]

Out[9]= {{1}, {2, 1, 1}, {3, 1, 2, 2}}
```

This list of positions is in the right form for MapAt.

```
In[10]:= MapAt[f, expr, %]
```

$$Out[10]= E^{1 + f[a]}\ c + \frac{b}{f[a]} + f[a]$$

This particular example could have been done more easily with a replacement rule.

```
In[11]:= expr /. a -> f[a]
```

$$Out[11]= E^{1 + f[a]}\ c + \frac{b}{f[a]} + f[a]$$

Another example of the use of Map[] and MapAt[] will be given in subsection 5.3.2.

■ 4.4.3 Apply

Apply[] implements a generalization of the usual notion of applying a function to an argument. When we speak of applying the function f to the expression *expr*, we mean to form the expression $f[expr]$ and to evaluate it. If we want to apply a function to several arguments, things get more complicated. Assume you have a list $l = \{e_1, e_2, \ldots, e_n\}$ of arguments and you want to compute $f[e_1, e_2, \ldots, e_n]$. Writing $f[l]$ would be wrong, it would pass the whole expression $\{e_1, e_2, \ldots, e_n\}$ as *one* argument to f. The expression Apply[f, l] does what we want. It forms the expression $f[e_1, e_2, \ldots, e_n]$. Looked at it from another point of view it replaces the head of l by f. Remember that $\{e_1, e_2, \ldots, e_n\}$ in internal form is just List[$e_1, e_2, \ldots, e_n$]. If we replace List by f we get $f[e_1, e_2, \ldots, e_n]$. If there is a special print form of an expression (as for lists or arithmetic operators), then this replacement of the head does not look so obvious.

The head of this expression is Plus.	In[1]:= **a + b + c**
	Out[1]= a + b + c
Replacing it by Times gives the product of the three terms in the sum above.	In[2]:= **Apply[Times, %]**
	Out[2]= a b c
This simple definition finds the average of a list of numbers.	In[3]:= **Average[l_List] := Plus @@ l / Length[l]**
	In[4]:= **Average[{1, 2, 3, 4, 5, 6}]**
	Out[4]= $\dfrac{7}{2}$
Of course, it also works for symbolic entries.	In[5]:= **Average[{a, b}]**
	Out[5]= $\dfrac{a + b}{2}$

■ 4.5 Application: The Platonic Solids

Mathematica comes with a package **Polyhedra.m** that defines functions for generating lists of polygons representing the five platonic solids (regular polyhedra) tetrahedron, cube, octahedron, dodecahedron, and icosahedron. It makes heavy use of Map[] and Apply[].

■ 4.5.1 Generating the Polygons

The idea behind **Polyhedra.m** is to describe each solid that we want to render by the coordinates of all of its vertices and by data that describes which vertices belong to each of the faces. For each solid, we give rules for the two functions Vertices[] and Faces[]. The function Vertices[*name*] gives the list of vertex coordinates for the solid with name *name*. Faces[*name*] gives the list of faces. The function Polyhedron[*name*] uses the information from Vertices[] and Faces[] to assemble a list of polygons in a Graphics3D[] object. Here is the relevant part of **Polyhedra.m**:

```
BeginPackage["Polyhedra`"]

Polyhedron::usage = "Polyhedron[name] gives a Graphics3D representing the named solid
    centered at the origin and with unit distance to the midpoints of the edges.
    Polyhedron[name, center, size] uses the given center and size."

Vertices::usage = "Vertices[name] gives a list of the vertex coordinates
    for the named solid."

Faces::usage = "Faces[name] gives a list of the faces for the named solid.
    Each face is a list of the numbers of the vertices that comprise that face."

Begin["`Private`"]

Polyhedron[ name_, pos_:{0.0, 0.0, 0.0}, scale_:1.0 ] :=
    Block[ {vertices = Vertices[name], faces = Faces[name]},
        Graphics3D[ Polygon /@ Map[scale # + pos &, (vertices[[#]]&) /@ faces, {2}] ]
    ]

Tetrahedron /: Faces[Tetrahedron] =
    {{1, 2, 3}, {1, 3, 4}, {1, 4, 2}, {2, 4, 3}}

Tetrahedron /: Vertices[Tetrahedron] = N[
    {{0, 0, 3^(1/2)}, {0, (2*2^(1/2)*3^(1/2))/3, -3^(1/2)/3},
     {-2^(1/2), -(2^(1/2)*3^(1/2))/3, -3^(1/2)/3},
     {2^(1/2), -(2^(1/2)*3^(1/2))/3, -3^(1/2)/3}} ]

End[]
EndPackage[]
```

Part of **Polyhedra.m**

How does `Polyhedron[]` work? The two local variables `vertices` and `faces` are set to the list of vertices and faces, respectively, of the solid we want. The code uses nested instances of `Map[]` written either in infix form as *f* /@ *expr* or in prefix form as `Map[`*f*, *expr*, *level*`]`. As usual, we can unwind the nested function applications to see what it does in detail.

The variable `vertices` is set to the list of vertex coordinates of the tetrahedron.

```
In[2]:= vertices = Vertices[Tetrahedron]
Out[2]= {{0., 0., 1.73205}, {0., 1.63299, -0.57735},
   {-1.41421, -0.816497, -0.57735},
   {1.41421, -0.816497, -0.57735}}
```

The tetrahedron has four faces. Each one is a triangle and so the list of faces is a list of four triples, each one specifying which three of the four vertices comprise that face.

```
In[3]:= faces = Faces[Tetrahedron]
Out[3]= {{1, 2, 3}, {1, 3, 4}, {1, 4, 2}, {2, 4, 3}}
```

We get the polygons by replacing the vertex numbers in the list of faces by the coordinates of the corresponding vertex. This statement gives us a list of faces, each of which is a list of the coordinates of the vertices.

```
In[4]:= (vertices[[#]]&) /@ faces
Out[4]= {{{0., 0., 1.73205}, {0., 1.63299, -0.57735},
   {-1.41421, -0.816497, -0.57735}},
  {{0., 0., 1.73205}, {-1.41421, -0.816497, -0.57735},
   {1.41421, -0.816497, -0.57735}},
  {{0., 0., 1.73205}, {1.41421, -0.816497, -0.57735},
   {0., 1.63299, -0.57735}},
  {{0., 1.63299, -0.57735},
   {1.41421, -0.816497, -0.57735},
   {-1.41421, -0.816497, -0.57735}}}
```

This statement applies the optional scaling and translation of the solid. The defaults don't change the coordinates (we multiply each coordinate by 1.0 and add 0.0).

```
In[5]:= Map[ 1.0 # + {0.0, 0.0, 0.0} &, %, {2} ];
```

Now we wrap the function `Polygon[]` around each face of the tetrahedron and make it into a graphics object.

```
In[6]:= Graphics3D[ Polygon /@ % ]
Out[6]= -Graphics3D-
```

And finally we get the picture.

In[7]:= **Show[%, Boxed -> False]**

Input line 4 above deserves some more explanation. The variable *faces* is a list of lists of vertex numbers. The first of these lists is {1, 2, 3} specifying that the first face of our tetrahedron consists of vertices 1, 2, and 3. We then apply the pure function `vertices[[#]]&` to each of the lists of faces. The first of the entries will therefore be `vertices[[{1, 2, 3}]]` (before evaluation). If the argument of a part extraction (*expr*`[[...]]`) is not a single number, but a list of numbers, then it returns the *subexpression* consisting of the elements specified. Since `vertices` is a list, the subexpression will again be a list, consisting of the coordinates of the vertices number 1, 2, and 3.

This generates a list of two polyhedra to be shown together in one picture.

In[8]:= **{Polyhedron[Icosahedron],**
                Polyhedron[Dodecahedron]}

Out[8]= **{-Graphics3D-, -Graphics3D-}**

The vertex coordinates were chosen so that the edges of two dual solids intersect in the middle. The dodecahedron and icosahedron are duals, as are the cube and octahedron.

In[9]:= **Show[%, Boxed -> False]**

■ 4.5.2 Manipulating Given Solids: Stellation

Once we have defined a solid in terms of its vertices and the list of faces, we can do more than just make pretty pictures. *Stellating* (or faceting) is one method of obtaining new solids from given ones. Each face is replaced by a pyramid with that face as its base and a new vertex, the apex of the pyramid, above the center of the face. The functional programming style makes it easy to replace each polygon in a graphics object by a pyramid. We use a rule that replaces each expression of the form Polygon[*list*] by a list of polygons, one for each face of the pyramid. Here is the code from **Polyhedra.m** for the command Stellate[*graphics*, *ratio*] that applies this transformation to each polygon found in *graphics*.

```
BeginPackage["Polyhedra`"]

⋮

Stellate::usage = "Stellate[-Graphics3D-, (stellation ratio:2)] replaces all polygons
    in the graphics by a pyramid. Stellation ratios smaller than 1 give concave figures."

Begin["`Private`"]

⋮

StellateFace[face_, k_] :=
    Block[ { i, apex, n = Length[face] } ,
        apex = N[ k Apply[Plus, face] / n ] ;
        Table[ Polygon[{apex, face[[i]], face[[ Mod[i, n] + 1 ]] }], {i, n} ]
    ]

Stellate[ graphics_, k_:2.0 ] :=
    graphics /. Polygon[x_] :> StellateFace[x, k] /; NumberQ[N[k]]
End[]
EndPackage[]
```

Faceting polygons

The externally visible command is Stellate[]. It consists of a single rule that replaces each polygon by the result of applying StellateFace[] to its argument. StellateFace[] takes as argument a list of vertices and first computes their center. If face is the list $\{v_1, v_2, \ldots, v_n\}$ in which each v_i is a list of three numbers (the coordinates of that vertex), then Apply[Plus, face] replaces the outer list by Plus and we get $v_1 + v_2 + \ldots + v_n$. Since Plus is listable and each of the v_i is a list of three numbers, they are added component by component and we get a single list of the coordinates as a result (Listability was explained in subsection 4.3.2). This list is then divided component by component by n, the number of vertices of that face. The result is a point that lies in the center of the original face. We have done nothing other than taken the average of the vertices. The apex of the pyramid is this point stretched by an arbitrary factor k.

The faces that make up this pyramid are triangles with two vertices along the base and the third one being the apex. The Table[] command picks out all sets of two consecutive vertices of the original face and combines them with the apex to a polygon. The Mod[] function used to pick out the second of the vertices causes the last point to be 1 and not n+1 which would lie beyond the end of the list. (When i runs from 1 to n, then mod$(i, n) + 1$ runs from 2 to n and then to 1.)

This gives a graphics object for the icosahedron.

```
In[2]:= Polyhedron[ Icosahedron ]

Out[2]= -Graphics3D-
```

This figure should look familiar. It is the logo of Wolfram Research, Inc.

```
In[3]:= Show[ Stellate[%] ]
```

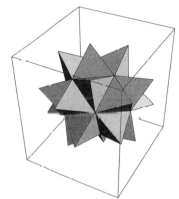

A stellation ratio smaller than 1 gives pyramids that point inward. The particular choice of $1/\sqrt{2}$ makes the new faces appear to be parts of pentagons with self-intersections. This solid is called a *great dodecahedron*.

```
In[4]:= Show[ Stellate[ %%, 1/Sqrt[2.0] ],
              Boxed -> False ]
```

`Stellate[]` works on any set of polygon. We can iterate it, for example, giving this doubly stellated icosahedron.

In[5]:= **Show[Stellate[%, 1.5]]**

This is a *small stellated dodecahedron*. The triangular faces are part of 12 pentagrams.

In[6]:= **Show[Stellate[Polyhedron[Dodecahedron], Sqrt[5.0]], Boxed -> False]**

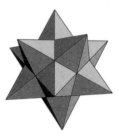

A different stellation ratio turns our logo into a *great stellated dodecahedron*. It too consists of 12 pentagrams.

In[7]:= **Show[Stellate[Polyhedron[Icosahedron], 3], Boxed -> False]**

■ 4.6 Operations on Lists and Matrices

Another important class of functional constructs manipulates the structure of expressions, mostly lists. `Transpose[]` and `Thread[]` rearrange the levels of nested lists. `RotateLeft[]` and `RotateRight[]` change the order of elements of lists. Finally, `Dot[]`, `Inner[]` and `Outer[]` combine elements of two nested expressions in various ways. We will look at all of these in turn.

■ 4.6.1 Rearranging Levels of Nested Expressions

The operation of transposing a matrix interchanges rows and columns of a matrix. We have used this at the beginning of Chapter 1 to get the columns of a matrix. Here is another useful application of `Transpose[]`.

```
In[1]:= n = Range[0, 10]
Out[1]= {0, 1, 2, 3, 4, 5, 6, 7, 8, 9, 10}
```

This finds the time it takes to expand the expression $(x + y + z + 1)^i$ for $i = 0, 1, ..., 10$.

```
In[2]:= Timing[ Expand[(x+y+z+1)^#] ][[1]]& /@ n
Out[2]= {0., 0., 0.0166667 Second, 0.0833333 Second,
    0.266667 Second, 0.516667 Second, 0.983333 Second,
    1.75 Second, 2.75 Second, 4.2 Second, 6.2 Second}
```

This combines the values of n and the timing information in a list.

```
In[3]:= { n, % /. Second -> 1 }
Out[3]= {{0, 1, 2, 3, 4, 5, 6, 7, 8, 9, 10},
    {0., 0., 0.0166667, 0.0833333, 0.266667, 0.516667,
    0.983333, 1.75, 2.75, 4.2, 6.2}}
```

Transposing this list builds pairs $\{i, time_i\}$.

```
In[4]:= Transpose[%]
Out[4]= {{0, 0.}, {1, 0.}, {2, 0.0166667},
    {3, 0.0833333}, {4, 0.266667}, {5, 0.516667},
    {6, 0.983333}, {7, 1.75}, {8, 2.75}, {9, 4.2},
    {10, 6.2}}
```

We need the data in this form to make a list plot of it.

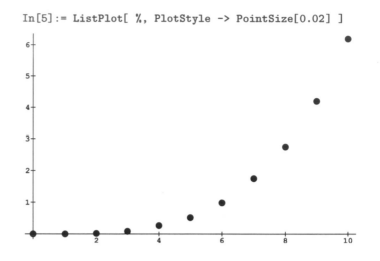

```
In[5]:= ListPlot[ %, PlotStyle -> PointSize[0.02] ]
```

Transpose[] takes an optional second argument that specifies which levels to interchange. If the levels in this level specification are not distinct, Transpose[] picks out diagonal elements in a matrix. This is useful, for example, to compute the *trace* of a matrix (the trace is the sum of the diagonal elements).

This definition computes the trace of a matrix, the sum of its diagonal elements.

```
In[6]:= Trace[m_?MatrixQ] := Plus @@ Transpose[m, {1, 1}]
```

Here is a 3 by 3 matrix.

```
In[7]:= {{a, b, c}, {d, e, f}, {g, h, i}} // MatrixForm

                                      a   b   c
                                      d   e   f
Out[7]//MatrixForm= g   h   i
```

This is its trace, the sum of its diagonal elements.

```
In[8]:= Trace[%]
Out[8]= a + e + i
```

The nested levels of an expression that is to be transposed need not be lists. Any head will do. The heads of the expressions must be same at all levels, however. Another operation, Thread[], is needed to interchange levels with different heads.

This expression cannot be transposed; the head of the outermost level is f, the inner level has head List.

```
In[9]:= f[{a, b}, {c, d}]
Out[9]= f[{a, b}, {c, d}]
```

Thread[] performs the operation.

```
In[10]:= Thread[%]
Out[10]= {f[a, c], f[b, d]}
```

Threading over lists can be made automatic by giving a function the attribute `Listable`.	`In[11]:= SetAttributes[g, Listable];`
Threading is done automatically.	`In[12]:= g[{a, b}, {c, d}]` `Out[12]= {g[a, c], g[b, d]}`

We will use `Thread[`*varlist* `->` *valuelist*`]` in section 7.3 to obtain a list of replacements of values for a list of variables.

■ 4.6.2 Rotating the Elements of an Expression

`RotateLeft[]` and `RotateRight[]` rotate the elements of an expression to the left or right. For an example of their use, we look at the auxiliary procedure `MakePolygons[]` in the package **ParametricPlot3D.m** that comes with every version of *Mathematica*.

The idea is similar to `MakeLines[]` in **ComplexMap.m** (see section 1.3). There we had a matrix of two-dimensional points that we connected to form horizontal and vertical lines. In **ParametricPlot3D.m**, we have a matrix of points in three dimensions that are to be connected as polygons. Here is an excerpt of **ParametricPlot3D.m**

```
BeginPackage["ParametricPlot3D`"]

ParametricPlot3D::usage =
    "ParametricPlot3D[{x,y,z}, {u,u0,u1,(du)}, {v,v0,v1,(dv)}, (options..)]
    plots a 3D parametric surface. Options are passed to Show[]"

Begin["`Private`"]

MakePolygons[vl_List] :=
    Block[{l = vl, l1 = Map[RotateLeft, vl], mesh},
        mesh = {l, l1, RotateLeft[l1], RotateLeft[l]};
        mesh = Map[Drop[#, -1]&, mesh, {1}];
        mesh = Map[Drop[#, -1]&, mesh, {2}];
        Polygon /@ Transpose[ Map[Flatten[#, 1]&, mesh] ]
    ]

Attributes[ParametricPlot3D] = {HoldFirst}

ParametricPlot3D[ fun_, ul:{_, u0_, u1_, du_}, vl:{_, v0_, v1_, dv_}, opts___] :=
    Show[Graphics3D[MakePolygons[Table[N[fun], ul, vl]]], opts] /;
        NumberQ[N[u0]] && NumberQ[N[u1]] && NumberQ[N[du]] &&
        NumberQ[N[v0]] && NumberQ[N[v1]] && NumberQ[N[dv]]

End[]
EndPackage[]
```

Part of **ParametricPlot3D.m**

As in `CartesianMap[]` we generate a two-dimensional table of values. The values are lists of three expressions that describe points in three dimensions. Any four neighboring points define a polygon in the surface to be drawn. If m_{ij} is the point in row i and column j, then the first polygon, for example, will consist of the points m_{11}, m_{12}, m_{22}, and m_{21}. We create four copies of m. In the second copy, we rotate each row, such that the point m_{12} will be in the top left corner. In the third copy, we also rotate the rows as a whole, such that the point m_{22} will be in the top left corner. In the fourth copy, the point m_{21} will be in the top left corner. To see how it works, we use symbols for the entries in the matrix of points, which makes it easier to track where they go under the rotations inside `MakePolygons[]`.

The parameter of `MakePolygons[]` is a matrix of points. The points are normally lists of numbers, but here we use symbols to make it easier to follow the computations.

```
In[1]:= (vl = { {a11, a12, a13}, {a21, a22, a23},
                {a31, a32, a33} }) // MatrixForm
              a11  a12  a13
              a21  a22  a23
Out[1]//MatrixForm= a31  a32  a33
```

`l1` has all its rows rotated one position to the left.

```
In[2]:= l = vl; l1 = Map[RotateLeft, vl]
Out[2]= {{a12, a13, a11}, {a22, a23, a21},
   {a32, a33, a31}}
```

This gives a list of four copies of `vl`, with rows rotated in different ways. The polygons can now be obtained by using the four points in the same position in the four copies. For example, `{a11, a12, a22, a21}` gives the first polygon.

```
In[3]:= mesh = {l, l1, RotateLeft[l1], RotateLeft[l]}
Out[3]= {{{a11, a12, a13}, {a21, a22, a23},
   {a31, a32, a33}}, {{a12, a13, a11}, {a22, a23, a21},
   {a32, a33, a31}}, {{a22, a23, a21}, {a32, a33, a31},
   {a12, a13, a11}}, {{a21, a22, a23}, {a31, a32, a33},
   {a11, a12, a13}}}
```

Since we don't want a closed surface, we have to throw away the last row ...

```
In[4]:= mesh = Map[Drop[#, -1]&, mesh, {1}];
```

... and the last column in each row.

```
In[5]:= mesh = Map[Drop[#, -1]&, mesh, {2}];
```

This flattens out the entries in each of the four copies of the original points.

```
In[6]:= Map[Flatten[#, 1]&, mesh]
Out[6]= {{a11, a12, a21, a22}, {a12, a13, a22, a23},
   {a22, a23, a32, a33}, {a21, a22, a31, a32}}
```

Here are the lists of vertices for the polygons.

```
In[7]:= Transpose[ % ]
Out[7]= {{a11, a12, a22, a21}, {a12, a13, a23, a22},
   {a21, a22, a32, a31}, {a22, a23, a33, a32}}
```

This surface is called a *catenoid*.

```
In[9]:= ParametricPlot3D[
            {Cosh[z] Cos[phi], Cosh[z] Sin[phi], z},
            {z, -2, 2}, {phi, 0, 2Pi}, Boxed -> False ]
```

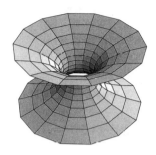

■ 4.6.3 Inner Products

The *inner product* or *dot product* is the usual matrix multiplication. You cannot use
the normal multiplication * for multiplying matrices. First, matrix multiplication is not
commutative; second, the normal multiplication is listable and is performed elementwise
and not according to the special formula for multiplying matrices. If you use matrices
and vectors with symbolic entries, it is easy to see what happens when you apply various
multiplication operators (See also sections 3.7.5 and 3.7.12 of the *Mathematica* book).

This generates a symbolic 3×3 matrix.

```
In[1]:= (m = Array[a, {3,3}]) // MatrixForm
                            a[1, 1]  a[1, 2]  a[1, 3]
                            a[2, 1]  a[2, 2]  a[2, 3]
Out[1]//MatrixForm= a[3, 1]  a[3, 2]  a[3, 3]
```

This is vector of length 3. We can use it as
either column- or row-vector. An explicit
distinction between the two is not necessary.

```
In[2]:= v = Array[b, 3]
Out[2]= {b[1], b[2], b[3]}
```

Here v is used a column vector. Left
multiplication by a matrix gives another column
vector as result.

```
In[3]:= m . v // ColumnForm
Out[3]= a[1, 1] b[1] + a[1, 2] b[2] + a[1, 3] b[3]
        a[2, 1] b[1] + a[2, 2] b[2] + a[2, 3] b[3]
        a[3, 1] b[1] + a[3, 2] b[2] + a[3, 3] b[3]
```

Here v is used as a row vector.

```
In[4]:= v . m
Out[4]= {a[1, 1] b[1] + a[2, 1] b[2] + a[3, 1] b[3],
        a[1, 2] b[1] + a[2, 2] b[2] + a[3, 2] b[3],
        a[1, 3] b[1] + a[2, 3] b[2] + a[3, 3] b[3]}
```

In contrast, this is what happens if we use the normal multiplication between a matrix and a vector.

```
In[5]:= m v
Out[5]= {{a[1, 1] b[1], a[1, 2] b[1], a[1, 3] b[1]},
         {a[2, 1] b[2], a[2, 2] b[2], a[2, 3] b[2]},
         {a[3, 1] b[3], a[3, 2] b[3], a[3, 3] b[3]}}
```

The dot product uses two operations to combine elements of its two arguments. It multiplies elements together and then adds up the resulting products. You can substitute your own functions for these two by using Inner[*multop*, m_1, m_2, *addop*], where m_1 and m_2 are the two matrices or vectors whose inner product is to be formed and *multop* is the function to use for the multiplication and *addop* is used for addition.

For an example, here is a very short definition for the divergence operator in Cartesian coordinates. Given an n-dimensional vector field $v = (e_1, e_2, \ldots, e_n)$ depending on n variables $(x_1, x_2, \ldots, x_n)$, then its divergence is given by the formula

$$\text{div } v = \frac{\partial e_1}{\partial x_1} + \frac{\partial e_2}{\partial x_2} + \cdots + \frac{\partial e_n}{\partial x_n}.$$

We can recognize this as a generalized inner product with differentiation replacing multiplication.

In a similar manner, we can implement the gradient function using application of the derivative operator to a list of variables. Given a scalar field $s(x_1, x_2, \ldots, x_n)$, then its gradient is the vector field

$$\text{grad } s = (\frac{\partial s}{\partial x_1}, \frac{\partial s}{\partial x_2}, \ldots, \frac{\partial s}{\partial x_n}).$$

Finally, the Laplacian of s can then be computed as $\nabla^2 s = \text{div grad } s$.

```
Div::usage = "Div[v, varlist] computes the divergence of
    the vector field v in Cartesian coordinates."

Grad::usage = "Grad[s, varlist] computes the gradient of s
    in Cartesian coordinates."

Laplacian::usage = "Laplacian[s, varlist] computes the laplacian of
    the scalar field s in Cartesian coordinates."

Div[v_List, var_List] := Inner[D, v, var, Plus]

Grad[s_, var_List] := D[s, #]& /@ var

Laplacian[s_, var_List] := Div[ Grad[s, var], var]
```

Part of **VectorCalculus.m**

Here is an example of its use with purely symbolic entries.

```
In[2]:= Laplacian[ v[x, y], {x, y} ]
```

$$Out[2]= v^{(0,2)}[x, y] + v^{(2,0)}[x, y]$$

Here is a standard gravitational or electrical force field.

```
In[3]:= e = {x/(x^2 + y^2 + z^2)^(3/2),
             y/(x^2 + y^2 + z^2)^(3/2),
             z/(x^2 + y^2 + z^2)^(3/2)};
```

This is its divergence.

```
In[4]:= Div[e, {x, y, z}]
```

$$Out[4]= \frac{-3\,x^2}{(x^2 + y^2 + z^2)^{5/2}} - \frac{3\,y^2}{(x^2 + y^2 + z^2)^{5/2}} -$$

$$\frac{3\,z^2}{(x^2 + y^2 + z^2)^{5/2}} + \frac{3}{(x^2 + y^2 + z^2)^{3/2}}$$

It needs some simplification to see that the result is correct.

```
In[5]:= Together[%]
```

```
Out[5]= 0
```

■ 4.6.4 Outer Products

The *outer product* is also described in section 3.7.12 of the *Mathematica* book. For an application let us look at the Jacobian matrix. Given a list of expressions $(e_1, e_2, \ldots, e_n)$ that describe functions of the variables $(x_1, x_2, \ldots, x_m)$, then the Jacobian is an $n \times m$ matrix of partial derivatives

$$\begin{pmatrix} \frac{\partial e_1}{\partial x_1} & \frac{\partial e_1}{\partial x_2} & \cdots & \frac{\partial e_1}{\partial x_m} \\ \frac{\partial e_2}{\partial x_1} & \frac{\partial e_2}{\partial x_2} & \cdots & \frac{\partial e_2}{\partial x_m} \\ \vdots & \vdots & & \vdots \\ \frac{\partial e_n}{\partial x_1} & \frac{\partial e_n}{\partial x_2} & \cdots & \frac{\partial e_n}{\partial x_m} \end{pmatrix}$$

This is an outer product with differentiation replacing multiplication. Here is the definition:

```
JacobianMatrix::usage = "JacobianMatrix[flist, varlist] computes the Jacobian of
    the functions flist w.r.t. the given variables."

JacobianMatrix[f_List, var_List] := Outer[D, f, var]
```

Definition of the Jacobian matrix using outer products

Here is a symbolic Jacobian.

```
In[6]:= JacobianMatrix[{f[x, y, z], g[x, y, z]},
                       {x, y, z}] // MatrixForm
Out[6]//MatrixForm=
   (1,0,0)            (0,1,0)            (0,0,1)
  f       [x, y, z]  f       [x, y, z]  f       [x, y, z]
   (1,0,0)            (0,1,0)            (0,0,1)
  g       [x, y, z]  g       [x, y, z]  g       [x, y, z]
```

Generalized outer products can also be used for generating combinatorial expressions. In one application we had to find the product of all terms of the form

$$x \pm \sqrt{p_1} \pm \sqrt{p_2} \cdots \pm \sqrt{p_n}$$

where the product ranges over all sign combinations and p_i denotes the ith prime number. Since for each term there are two choices of the sign, there are a total of 2^n terms. This can be generated as an outer sum.

This outer sum generates all the terms in a multi-dimensional tensor.

```
In[1]:= Outer[Plus, {x}, {-Sqrt[2],  Sqrt[2]},
                    {-Sqrt[3], Sqrt[3]}, {-Sqrt[5], Sqrt[5]}]
Out[1]= {{{{-Sqrt[2] - Sqrt[3] - Sqrt[5] + x,
      -Sqrt[2] - Sqrt[3] + Sqrt[5] + x},
    {-Sqrt[2] + Sqrt[3] - Sqrt[5] + x,
      -Sqrt[2] + Sqrt[3] + Sqrt[5] + x}},
   {{Sqrt[2] - Sqrt[3] - Sqrt[5] + x,
     Sqrt[2] - Sqrt[3] + Sqrt[5] + x},
    {Sqrt[2] + Sqrt[3] - Sqrt[5] + x,
     Sqrt[2] + Sqrt[3] + Sqrt[5] + x}}}}
```

We flatten out the tensor and replace the outermost list with a product to get our desired result.

```
In[2]:= Times @@ Flatten[ % ]
Out[2]= (-Sqrt[2] - Sqrt[3] - Sqrt[5] + x)
  (Sqrt[2] - Sqrt[3] - Sqrt[5] + x)
  (-Sqrt[2] + Sqrt[3] - Sqrt[5] + x)
  (Sqrt[2] + Sqrt[3] - Sqrt[5] + x)
  (-Sqrt[2] - Sqrt[3] + Sqrt[5] + x)
  (Sqrt[2] - Sqrt[3] + Sqrt[5] + x)
  (-Sqrt[2] + Sqrt[3] + Sqrt[5] + x)
  (Sqrt[2] + Sqrt[3] + Sqrt[5] + x)
```

These terms are polynomials with integer coefficients. They are called the *Swinnerton-Dyer polynomials*.

```
In[3]:= Expand[%]
                      2          4        6    8
Out[3]= 576 - 960 x  + 352 x  - 40 x  + x
```

Here is a definition that generates the Swinnerton-Dyer polynomials for any n:

```
SwinnertonDyerP::usage =
    "SwinnertonDyerP[n, var] gives the minimal polynomial
    of the sum of the squareroots of the first n primes."

Begin["`Private`"]

SwinnertonDyerP[n_Integer?Positive, x_] :=
    Block[{arglist, poly, i},
        arglist = Outer[Times, Table[Sqrt[Prime[i]], {i, n}], {-1, 1}];
        poly = Outer[Plus, {x}, Sequence @@ arglist];
        Expand[Times @@ Flatten[poly]]
    ]

End[]
Null
```

SwinnertonDyer.m: generating the Swinnerton-Dyer polynomials

Here is how it works:

In this example, we want to compute SwinnertonDyerP[3, y] and so we assign the parameters to the pattern names.

```
In[4]:= n = 3; x = y;
```

This generates a list of pairs of the positive and negative square roots of the first three primes.

```
In[5]:= arglist = Outer[Times,
            Table[Sqrt[Prime[i]], {i, n}], {-1, 1}]

Out[5]= {{-Sqrt[2], Sqrt[2]}, {-Sqrt[3], Sqrt[3]},
    {-Sqrt[5], Sqrt[5]}}
```

The outer list is replaced by a sequence. Each of the pairs of square roots will therefore occupy one position in the argument list of the next call to Outer.

```
In[6]:= Sequence @@ arglist
Out[6]= ({-Sqrt[2], Sqrt[2]}, {-Sqrt[3], Sqrt[3]},
    {-Sqrt[5], Sqrt[5]})
```

The call to Outer[] is now exactly the same as in the example above (see input line 1). We suppress the lengthy result.

```
In[7]:= poly = Outer[Plus, {x}, %];
```

We get the same result as before.

```
In[8]:= Expand[Times @@ Flatten[poly]]
Out[8]= 576 - 960 y^2 + 352 y^4 - 40 y^6 + y^8
```

Here is the next of these polynomials. It has degree 16.

```
In[9]:= SwinnertonDyerP[4, z]
Out[9]= 46225 - 5596840 z^2 + 13950764 z^4 - 7453176 z^6 +
    1513334 z^8 - 141912 z^10 + 6476 z^12 - 136 z^14 + z^16
```

Chapter 5

Evaluation

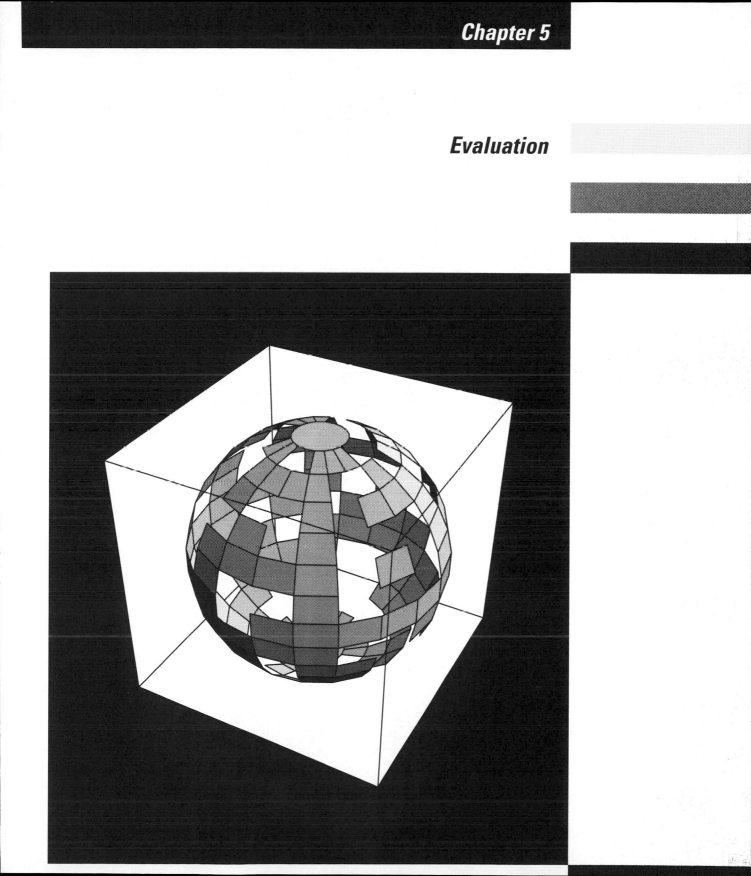

In this chapter we look at various aspects of *Mathematica*'s way of evaluating expressions. The usual way of evaluating expressions is quite straightforward, but there are many exceptions and subtle points. Some knowledge of how expressions are evaluated is necessary to write good programs in *Mathematica*.

Section 1 looks at how rules are applied. There is an important difference between rules in *Mathematica* and procedures in traditional languages. It concerns the way pattern names are used inside the body of a rule. Clever use of this setup allows you to write functions that return other functions as results. This section shows you how to do this.

Pure functions are the topic of section 2. A concept not available in most other languages, they are a very useful tool in *Mathematica*. We use them heavily in this book.

There are cases when you do not want to evaluate an argument of a function that is normally evaluated. Section 3 describes how to achieve this and what it is good for. On the other hand, there are a few built-in functions that do not evaluate their arguments. You can force evaluation in this case and section 4 treats this topic. It is in a way the opposite of the previous section.

In section 5 we look at an application of the ideas treated so far. We write a function that defines other functions for us—all under program control.

Assigning values to variables is a simple concept but (no surprise), in *Mathematica* this idea has advanced uses. One of the examples presented assumes some familiarity with the operating system UNIX. You may safely skip this subsection, if this topic does not interest you.

Finally, section 7 treats yet another subtlety encountered when defining rules. It has to do with the way the left side of a definition is evaluated. If you try to make a definition and you get strange error messages or the rule does not match the intended cases, try looking in this section for a possible cause of the problem.

About the illustration overleaf:

The Sphere[] command from the package **Shapes.m** creates polygons that approximate a sphere. In this picture we have randomly removed half of these polygons.

```
<< Graphics/Shapes.m
Show[ Graphics3D[Select[Sphere[][[1]], Random[]>0.5&]] ]
```

■ 5.1 Evaluation of the Body of a Rule

■ 5.1.1 Pattern Names and Local Variables

The *Mathematica* book contains a discussion of the way the names of patterns are used inside the body of a rule (subsection 2.5.10). Let us look at the issues involved in some more detail. Assume you give a rule like

$$\text{rule} = f[x_-] :> body$$

or a definition like

$$f[x_-] := body$$

and then evaluate f[*expr*] /. rule or f[*expr*]. The expression f[*expr*] matches the pattern f[x_] with *expr* matching x. Evaluation then proceeds with the right side of the rule, the expression *body*. Before it is evaluated, all occurrences of x are replaced by *expr*, much in the same way as if you had given the expression

$$\text{Release}[\text{ Hold}[body] \text{ /. } x \rightarrow expr \text{]}.$$

Written in this way, the Hold[] is necessary since normally the left side of a replacement is evaluated before the replacement is performed, but the body of a rule is not evaluated before the replacement. A consequence of this is that names of patterns cannot be used as local variables. It is not possible to assign to them.

This definition tries to compute the numerical value of x and then return its square root.	In[1]:= f[x_] := (x = N[x]; Sqrt[x])
An error message results.	In[2]:= f[2] Set::raw: Cannot assign to raw object 2. Out[2]= Sqrt[2]
Simulating the evaluation of the body of the definition makes it clear what happened. After the substitution we would try to assign to the number 2.	In[3]:= Hold[(x = N[x]; Sqrt[x])] /. x -> 2 Out[3]= Hold[2 = N[2]; Sqrt[2]]

If you want to use pattern names as local variables in the same way as procedure parameters are used in the programming languages C and Pascal, for example, you can do it by using an initialized local variable.

```
f[x_] :=
    Block[{xx = x},
        xx = N[xx];
        Sqrt[xx]
    ]
```

How to use pattern names as local variables

■ 5.1.2 Functions that Return Functions

The mechanism explained in the previous subsection is important if we want to write functions that return other functions. Let us write a function MultByN[n] that returns a pure function that multiplies its argument by n.

Here we use the fact that replacement of the argument value is done everywhere on the right side, even inside pure functions, whose body is not evaluated.

```
In[1]:= MultByN[n_] := Function[x, x n]
```

m5 is a function that multiplies its argument by 5.

```
In[2]:= m5 = MultByN[5]

Out[2]= Function[x, x 5]
```

Applied to the argument 7 we get 35.

```
In[3]:= m5[7]
Out[3]= 35
```

A function MultByN2[n] that returns a function that multiplies its argument by n^2 is harder to write.

Here is our first attempt.

```
In[4]:= MultByN2[n_] :=
            Block[{n2 = n^2},
                Function[x, x n2]
            ]
```

That's not what we wanted. n2 has not been replaced by its value, since the body of the pure function was not evaluated inside MultByN2[].

```
In[5]:= MultByN2[5]

Out[5]= Function[x, x n2]
```

We have to use substitution.

```
In[6]:= MultByN2[n_] :=
            Block[{n2},
                Function[x, x n2] /. n2 -> n^2
            ]
```

Now it works.

```
In[7]:= MultByN2[5]
Out[7]= Function[x, x 25]
```

$7 \cdot 5^2 = 175$.

```
In[8]:= %[7]
Out[8]= 175
```

The operation of differentiation is also of this kind. It takes a function as argument and returns its derivative, again a function.

Here is an example function definition.

```
In[1]:= f[x_] := 1 + x^2
```

This computes the first derivative of f. Since we do not have a a name for this function, it is returned as a pure function.

```
In[2]:= f'
Out[2]= 2 #1 &
```

Applying it to an argument gives the expected result.

```
In[3]:= %[z]
Out[3]= 2 z
```

The argument of the derivative operator can of course itself be a pure function. This is the same function as f above and so is, of course, its derivative.

```
In[4]:= (1 + #^2)&'
Out[4]= 2 #1 &
```

The derivative of a pure function is computed by differentiating its body with respect to its variable, in this case computing D[y^x, x]. The result is, however, not simplified because the body of a pure function must not be evaluated.

```
In[5]:= Function[x, y^x]'
                          x
Out[5]= Function[x, y  1 Log[y]]
```

As soon as the function is applied to an argument, it is fully evaluated and simplified.

```
In[6]:= %[z]
              z
Out[6]= y  Log[y]
```

■ 5.2 Pure Functions

■ 5.2.1 Introduction

We have already encountered *pure functions* several times. They are discussed in sections 2.1.12 and 2.1.13 of the *Mathematica* book. Given an expression *expr* involving some variable x, you can think of it as describing a function with that variable being the argument. If you need to refer to that function by name, you can either give a rule for a symbol f in the form

$$f[x\_] := expr$$

or you can use the object

$$\texttt{Function[}x, expr\texttt{]}.$$

Either f or Function[x, *expr*] is a name for that function and you can use the two interchangeably. To apply them to an argument y, you write f[y] in the familiar way or Function[x, *expr*][y], the latter perhaps looking rather strange at first.

This defines f to be the function with $f(x) = 1 + x^2$.	`In[1]:= f[x_] := 1 + x^2`

g is set to be a pure function that describes the same function as f.	`In[2]:= g = Function[x, 1 + x^2]`
	
	`Out[2]= Function[x, 1 + x ]`

They can be used in the same way.	`In[3]:= {f[3], g[3]}`
	
	`Out[3]= {10, 10}`

There are cases where a named function, like f above, cannot be used. Examples are functions that return functions as their result, which was discussed in subsection 5.1.2.

■ 5.2.2 Theoretical Properties

The study of formal properties of systems of pure functions is a branch of mathematics called λ-*calculus*. Three of the theorems of λ-calculus are of particular interest for *Mathematica*'s pure functions. In λ-calculus, they are called λ-conversion, α-conversion, and η-conversion.

The first theorem (λ-conversion) states how pure functions are applied to arguments. This is done by replacing every occurrence of the formal variable in the body of the pure function by the argument. In *Mathematica* notation, the expression

$$\texttt{Function[}x, body\texttt{][}arg\texttt{]}$$

is evaluated as

$$\texttt{Release[Hold[}body\texttt{] /. Literal[}x\texttt{] -> }arg\texttt{]}.$$

This is essentially the same as

$$body \; /. \; x \; \text{->} \; arg$$

except that *body* and *x* are not evaluated before the rule is applied. Note that this is similar to how the body of a rule is evaluated as we have seen in section 5.1.

We set x to 0 so that we would later detect any attempt to evaluate the body of a pure function involving *x* before the argument is substituted for *x*.

```
In[1]:= x=0
Out[1]= 0
```

This gives what we expect and is not influenced by the global value for x.

```
In[2]:= Function[x, 1/x][4]

         1
Out[2]= -
         4
```

This expression is equivalent. It shows how pure functions are applied to arguments.

```
In[3]:= Release[ Hold[1/x] /. Literal[x] -> 4 ]

         1
Out[3]= -
         4
```

The second theorem (α-conversion) states essentially that the name of the formal variable does not matter. If we have a pure function Function[x, *expr*], replacing the variable x by y everywhere in *expr* describes the exact same function.

Here is a sample pure function.

```
In[1]:= Function[x, Sqrt[x] + Sin[x] + E^(2x)]

                                            2 x
Out[1]= Function[x, Sqrt[x] + Sin[x] + E   ]
```

This expression describes the same function.

```
In[2]:= % /. x -> y

                                            2 y
Out[2]= Function[y, Sqrt[y] + Sin[y] + E   ]
```

As we have seen, the expression Function[x, *body*][*arg*] is evaluated by substituting the value of the argument *arg* for every occurrence of the variable x in *body*. Obviously the name of the variable does not matter. This is only true, however, if y did not occur in *body* before the replacement.

The third theorem (η-conversion) states that a pure function having the form Function[*var*, *f*[*var*]] is the same function as simply *f* itself.

This is merely a complicated way of writing the sine function.

```
In[3]:= Function[x, Sin[x]]
Out[3]= Function[x, Sin[x]]
```

Applying it to an argument gives the same as just applying Sin itself to that argument.

```
In[4]:= %[z]
Out[4]= Sin[z]
```

■ 5.2.3 Short Forms of Pure Functions

Since the names of the variables in a pure function do not matter, *Mathematica* can provide special symbols for denoting these variables. The expressions #1, #2, ... are used for the first, second, etc. variable. Their internal form is Slot[*i*]. If you use these, you can leave out the first argument of Function[] that serves to declare the names of the variables. A convenient abbreviation for #1 is #. We have used this quite frequently in this book.

The variables need not occur at all in the body of a pure function. The expression 1& is a constant function that—independent of the values of its arguments—always returns 1.

It is an error to use fewer than the declared number of arguments.

```
In[1]:= Function[{a, b}, a][1]

Function::count:
    Too many parameters {a, b} to be filled from
    Function[{a, b}, a][1].

Out[1]= 1
```

Any excess arguments are simply ignored.

```
In[2]:= Function[{a, b}, a][1, 2, 3]

Out[2]= 1
```

This defines a function constant[*val*] whose value is a constant pure function that always returns *val*.

```
In[3]:= constant[x_] := x&
```

This defines k7 to be a function that always returns 7.

```
In[4]:= k7 = constant[7]

Out[4]= 7 &
```

Whatever the argument, it returns 7.

```
In[5]:= k7[666]

Out[5]= 7
```

■ 5.3 Preventing Evaluation of Arguments of Functions

■ 5.3.1 Functions that do not Evaluate their Arguments

Normally, functions evaluate their arguments before any rules for that function take effect. There are, however, special cases. We have already seen such examples, namely all the iterators. The prototypical function that does not evaluate its argument is Hold[*expr*]. It does nothing, but its presence prevents *expr* from being evaluated. For more on evaluation, see subsection 2.2.19 of the *Mathematica* book. If you set the attribute HoldAll for a function, it will not evaluate its arguments. If you give a rule like f[e_] := *body*, then the unevaluated argument is substituted for every occurrence of e in the body of the rule. Unless such an occurrence is again in a function that does not evaluate its arguments, it will be evaluated there, inside your function.

As an example of this, here is a function PrintTime[] that prints the time it takes to evaluate its argument and then returns the result of that evaluation. It is clear that it must not evaluate its argument before it passes it to the built-in function Timing[] that does the time measurement.

```
PrintTime::usage = "PrintTime[expr] prints the time it takes
    to evaluate expr and returns the result of the evaluation."

Begin["`Private`"]

SetAttributes[PrintTime, HoldAll]

PrintTime[expr_] :=
    Block[{`timing},
        timing = Timing[expr];
        Print[ timing[[1]] ];
        timing[[2]]
    ]

End[]
Null
```

PrintTime.m

The unevaluated argument is passed to Timing[] inside PrintTime[]. Timing[] itself does not evaluate its argument either. It is evaluated inside the built-in code of Timing[]. Timing[] returns a list {*time*, *result*}. We print the time and return the result.

The printing of the time is a side effect that does not interfere with the normal course of the computation.

```
In[2]:= PrintTime[ Factor[x^10 - y^10 ]]
0.0833333 Second
```

$$Out[2]= (x - y) (x + y) (x^4 - x^3 y + x^2 y^2 - x y^3 + y^4)$$
$$(x^4 + x^3 y + x^2 y^2 + x y^3 + y^4)$$

`Timing[]` returns this list and so it can disturb the normal flow of computation, while `PrintTime[]` is almost invisible. Timing information is not very reproducible and so the two times might be different.

```
In[3]:= Timing[ Factor[x^10 - y^10 ] ]
Out[3]= {0.116667 Second,
```
$$(x - y) (x + y) (x^4 - x^3 y + x^2 y^2 - x y^3 + y^4)$$
$$(x^4 + x^3 y + x^2 y^2 + x y^3 + y^4)\}$$

We will see a better way of keeping track of the timing for each command in section 8.2.2.

■ 5.3.2 Application: Extracting Parts of Held Expressions

Let us address the problem of finding the structure of an expression inside Hold[] without evaluating it. We can, of course, extract the expression inside Hold[*expr*] by either Hold[*expr*][[1]] or Release[Hold[*expr*]]. But if we use this inside Length[], for example (to find the the length of *expr*), then it would be evaluated, since Length[] does evaluate its argument. The solution is to use MapAt[] to wrap Hold[] around each element in *expr* (including the head) and then get rid of the outer Hold[]. The expression is now completely "frozen" and we can find its length, for example.

This expression, when given a chance to evaluate, would immediately turn into 26.

```
In[1]:= expr = Hold[ 0 1 + 2 3 + 4 5 ]
Out[1]= Hold[0 1 + 2 3 + 4 5]
```

This takes care of the head, which would not be necessary here, since it is a symbol without a value.

```
In[2]:= expr = MapAt[ Hold, expr, {1, 0} ]
Out[2]= Hold[Hold[Plus][0 1, 2 3, 4 5]]
```

Map[] maps Hold at each element of our original expression.

```
In[3]:= expr = Map[ Hold, expr, {2} ]
Out[3]= Hold[Hold[Plus][Hold[0 1], Hold[2 3], Hold[4 5]]]
```

This gets rid of the outer Hold[].

```
In[4]:= expr = expr[[1]]
Out[4]= Hold[Plus][Hold[0 1], Hold[2 3], Hold[4 5]]
```

Here is its length without evaluating it.

```
In[5]:= Length[expr]
Out[5]= 3
```

This extracts its second element without evaluating it.

```
In[6]:= expr[[2]]
Out[6]= Hold[2 3]
```

The function `WrapHold[]` implements these steps.

```
WrapHold::usage = "WrapHold[expr] wraps Hold[] around the head
    and the elements of expr without evaluating them."

SetAttributes[WrapHold, HoldAll]

WrapHold[expr_] :=
    Map[ Hold, MapAt[Hold, Hold[expr], {1, 0}], {2}] [[1]]
```

WrapHold.m: Wrap `Hold[]` around all parts of an expression

You wouldn't want any of these expressions evaluated (for different reasons). This example makes a good test of `WrapHold[]` since any mistake would not go unnoticed.

`In[2]:= WrapHold[ Exit[Quit[], 1/0, 3^10^10] ]`

$$Out[2]= Hold[Exit][Hold[Quit[]], Hold[\frac{1}{0}], Hold[3^{10^{10}}]]$$

The functions `Edit[]`, `EditIn[]` and `EditDef[]` that should be available in most versions of *Mathematica* (they are not built-in) make heavy use of ideas like this.

■ 5.4 Forcing Evaluation of Arguments of Functions

■ 5.4.1 Evaluating Arguments that are Normally not Evaluated

If a function has the attribute `HoldAll` then its arguments are used inside the function body without prior evaluation, as we have just seen. Evaluation can be forced in such a case by using `Release[]`.

This measures the time it takes to sum the integers from 1 to 1000. The `Sum[]` is evaluated only inside the function `Timing[]`.

```
In[1]:= Timing[ Sum[i, {i, 1000}] ]
Out[1]= {1.83333 Second, 500500}
```

In this case, the sum is evaluated before it is passed to `Timing[]`. `Timing[]` then receives the integer 500500 as argument and it takes almost no time to evaluate it again.

```
In[2]:= Timing[ Release[Sum[i, {i, 1000}]] ]
Out[2]= {0., 500500}
```

Sometimes `Release[]` is necessary. The command `Plot[]`, for example, does not evaluate its first argument. It looks at the unevaluated argument to see whether it is a list, in which case it prepares to plot several functions in one picture. If it is not a list, it assumes it is a single function. If you have a list of expressions stored as a value of some variable and you want to use it in `Plot[]` to plot all of these expressions in one picture, evaluating it (to a list of expressions) before `Plot[]` sees it is essential.

This generates a list of the first 10 Chebyshev polynomials.

```
In[1]:= Table[ ChebyshevT[i, x], {i, 1, 10} ]
```

$$Out[1]= \{x,\ -1 + 2\ x^2,\ -3\ x + 4\ x^3,\ 1 - 8\ x^2 + 8\ x^4,$$
$$5\ x - 20\ x^3 + 16\ x^5,\ -1 + 18\ x^2 - 48\ x^4 + 32\ x^6,$$
$$-7\ x + 56\ x^3 - 112\ x^5 + 64\ x^7,$$
$$1 - 32\ x^2 + 160\ x^4 - 256\ x^6 + 128\ x^8,$$
$$9\ x - 120\ x^3 + 432\ x^5 - 576\ x^7 + 256\ x^9,$$
$$-1 + 50\ x^2 - 400\ x^4 + 1120\ x^6 - 1280\ x^8 + 512\ x^{10}\ \}$$

Here is a picture of them. It would not have worked in the form `Plot[%, {x, -1, 1}]`.

`In[2]:= Plot[ Release[%], {x, -1, 1} ]`

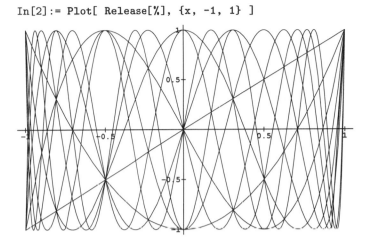

Sometimes `Release[]` is merely a matter of efficiency. For an example, let us look again at `SphericalPlot3D[]` (see subsection 2.2.4). `SphericalPlot3D[]` does not evaluate its first argument but passes it on to `ParametricPlot3D[]` which does not evaluate it either. It is finally evaluated inside a `Table[]` command. If we want to generate a spherical plot of the function `Abs[SphericalHarmonicY[3, 1, theta, phi]]`, then that function is evaluated over and over again. `SphericalHarmonicY[]` does evaluate to a polynomial in trigonometric functions, however, and in this form would take much less time to evaluate numerically.

Generating this plot takes only about 60% of the time that it would take without `Release[]`.

```
In[2]:= SphericalPlot3D[ Release[
            Abs[SphericalHarmonicY[3, 1, theta, phi]]],
         {theta, 0, Pi}, {phi, 0, 2Pi} ]
```

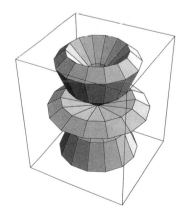

This is the expression into which
SphericalHarmonicY[] evaluates.

```
In[3]:= SphericalHarmonicY[3, 1, theta, phi]

              I phi                        2
        -(Sqrt[7] E     (-3 + 15 Cos[theta] ) Sin[theta])
Out[3]= ──────────────────────────────────────────────────
                      8 Sqrt[3] Sqrt[Pi]
```

By noting that the absolute value of it is
independent of phi and by already performing
a numerical approximation, we can plot it even
faster.

```
In[4]:= N[% /. phi -> 0]

                                          2
Out[4]= 0.107727 ( 3. + 15. Cos[theta] ) Sin[theta]
```

We use the saved time to increase the plot
resolution.

```
In[5]:= SphericalPlot3D[ Release[Abs[%]],
          {theta, 0, Pi, Pi/24}, {phi, 0, 2Pi} ]
```

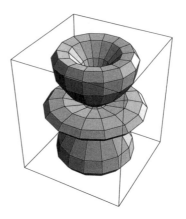

■ 5.4.2 A Second Use of Release

Release[] has a second use. It removes any occurrences of Hold[] in an expression.
As such, it can only be used at top level and not inside a definition. The reason is that the
first use explained in the previous subsection takes precedence and Release[] is then
removed from the definition. This is an unfortunate feature that is likely to be remedied
in a future version of *Mathematica*.

Before defining a rule for the function test[],
we give it the attribute HoldAll.

```
In[1]:= Attributes[test] = {HoldAll}
```

With this rule, we try to remove the Hold[]
that prevents the unevaluated argument from
being evaluated inside test[].

```
In[2]:= test[x_] := Release[Hold[x]]
```

Release[] does not seem to work.

```
In[3]:= test[1+1]
Out[3]= Hold[1 + 1]
```

In fact, it is no longer there at all! It was used to force evaluation of the second argument of `SetDelayed[]` (the internal form of `:=`) when `test[]` was defined. There it did not have any effect.

```
In[4]:= ?test

test
Attributes[test] = {HoldAll}
test/: test[x_] := Hold[x]
```

We can outwit the evaluator by applying `Release[]` to our expression later, when the rule will be used.

```
In[4]:= test[x_] := Release @@ {Hold[x]}
```

This is the result we wanted.

```
In[5]:= test[1+1]
Out[5]= 2
```

Another instance where `Release[]` does something unexpected is inside a compound expression (statements separated by semicolons). The compound expression *expr₁* ; *expr₂* has internal form `CompoundExpression[`*expr₁* `,` *expr₂*`]`. The expression *expr₁* ; `Release[`*expr₂*`]` is therefore merely a special case of using `Release[]` to evaluate arguments of a function that does not evaluate its arguments (see subsection 5.4.1).

The result of this statement is rather surprising.

```
In[1]:= Block[{local = 3}, local = 5; Release[local]]
Out[1]= 3
```

Here is what happens: `Block[]` does not evaluate its arguments. It first assigns any values given for the declared variables and so it sets `local` to 3. Now it evaluates its second argument `local = 5; Release[local]` which is of the form just mentioned. `CompoundExpression[]` does not evaluate its arguments but the second one has `Release[]` wrapped around it. It is therefore evaluated and its value is of course 3. The expression now reads `local = 5; 3`. `CompoundExpression[]` now evaluates its arguments in turn, returning the value of the last one.

■ 5.5 Application: Programmed Definition of Functions

In subsection 5.1.2 we have seen a way to define functions that return functions as their values. In this section, we want to look at functions that define rules for another function. For an example, assume we need to set up a variety of *step functions*. A step function $f(x)$ is a function that takes on just two values, one value a if the argument x is less than some value x_0 and another value b if the argument is greater than x_0. In *Mathematica*, this can easily be defined as follows:

```
f[x_] := a /; x <= x0
f[x_] := b /; x >  x0
```

Defining a step function

Now we want to write a function StepFunction[f, a, x_0, b] that sets up such a definition for f. This is easy:

```
StepFunction[f_Symbol, a_, x0_, b_] := (
    f[x_] := a /; x <= x0;
    f[x_] := b /; x >  x0
)
```

The first version of StepFunction[]

This works because the values of the pattern variables f, a, x0, and b are *substituted* in the body of the rule. Evaluating the body globally defines the rules for f, as if we had typed them in. The name of the variable used in the rules defined for f does not matter at all. Note that the parentheses are necessary.

This defines the *sign* function. (It is already built in as Sign[].)

```
In[2]:= StepFunction[Signum, -1, 0, 1]
```

The rules have indeed been defined.

```
In[3]:= ?Signum
Signum
Signum/: Signum[x_] := -1 /; x <= 0

Signum/: Signum[x_] := 1 /; x > 0
```

Here is a plot of it.   In[3]:= Plot[Signum[x], {x, -1, 1}]

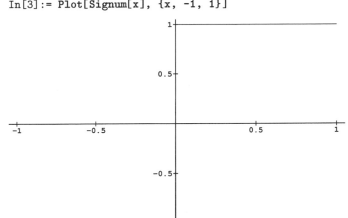

   In this example, the right side and the condition of the rules involved only the parameters of the function StepFunction[] itself. They were therefore substituted correctly everywhere inside the body of StepFunction[]. If the right side or the condition is more complicated and has to be computed first, then this would not work. The definitions set up for f are not evaluated inside StepFunction[]. We have to use substitution to get the computed values inside these definitions in the same way as we did in subsection 5.1.2. Another approach is to write an auxiliary function MakeRuleConditional[] that defines a rule. We can then pass the parts of the rule as parameters and again use substitution.

   The package **MakeFunctions.m** implements this auxiliary function and uses it to define the two functions StepFunction[] and LinearFunction[].

```
BeginPackage["MakeFunctions`"]

StepFunction::usage = "StepFunction[f, a, x0, b] defines rules for f
    such that f[x] = a for x <= x0, f[x] = b for x > x0."

LinearFunction::usage = "LinearFunction[f, a, x0, x1, b] defines rules for f
    such that f[x] = a for x <= x0, f[x] = b for x >= x1 and
    f increases linearly from a to b between x0 and x1 "

MakeRule::usage = "MakeRule[f, x, rhs]
    globally defines the rule f[x_] := rhs."

MakeRuleConditional::usage = "MakeRuleConditional[f, x, rhs, cond]
    globally defines the rule f[x_] := rhs /; cond."

Begin["`Private`"]

`x    (* used for the dummy variable in the functions defined *)

SetAttributes[MakeRule, HoldAll]

MakeRule[f_Symbol, var_Symbol, rhs_] :=
    f[var_] := rhs

SetAttributes[MakeRuleConditional, HoldAll]

MakeRuleConditional[f_Symbol, var_Symbol, rhs_, condition_] :=
    (f[var_] := rhs /; condition)

StepFunction[f_Symbol, a_, x0_, b_] := (
    MakeRuleConditional[f, x, a, x <= x0];
    MakeRuleConditional[f, x, b, x > x0]
    )

LinearFunction[f_Symbol, a_, x0_, x1_, b_] :=
    Block[{slope = (b-a)/(x1-x0)},
        MakeRuleConditional[f, x, a, x <= x0];
        MakeRuleConditional[f, x, Release[a + (x-x0) slope], x0 < x < x1];
        MakeRuleConditional[f, x, b, x >= x1]
    ]
End[]

Protect[StepFunction, LinearFunction, MakeRule, MakeRuleConditional]

EndPackage[]
```

MakeFunctions.m: Utilities for defining functions

LinearFunction[f, a, x_0, x_1, b] defines $f(x)$ to be a function whose values are a for $x \leq x_0$, b for $x \geq x_1$ and linearly increases from a to b between x_0 and x_1.

Mathematica warns when the right side of a definition contains a pattern. This happens in MakeRule[] and MakeRuleConditional[]. In this case, this is all right, since the right side is another definition.

```
In[1]:= << MakeFunctions.m
Pattern::rhs:
    var_ appears on right-hand side of rule; omit _ ?
Pattern::rhs:
    var_ appears on right-hand side of rule; omit _ ?
```

This defines rules for g.

```
In[2]:= LinearFunction[g, -1, -1, 1, 1]
```

And here is a picture of g.

```
In[3]:= Plot[g[y], {y, -2, 2}]
```

This code contains a few subtleties:

- The functions MakeRule[] and MakeRuleConditional[] do not evaluate their arguments. This is desirable so that unevaluated right sides and conditions can be specified. If these arguments contain local variables, however, they must be evaluated because the definition set up for f will eventually be used outside the scope of these local variables. Therefore, the formula a + (x-x0) slope used in LinearFunction[] is put inside Release[].

- The parentheses around the body of MakeRuleConditional[] are necessary. If they had been omitted, the condition following /; would be associated with MakeRuleConditional[] itself and not with the rule for f that it defines.

The error messages that are printed when **MakeFunctions.m** is read in can be suppressed by surrounding the definitions of MakeRule[] and MakeRuleConditional[] with Off[Pattern::rhs] ... On[Pattern::rhs]. Normally, turning off error messages to make your packages read in without errors is, of course, a bad idea. In this case, it would be rather awkward to fix and the error message does not imply an error but merely a warning that is not appropriate in this case.

■ 5.6 Assigning Values to Expressions

■ 5.6.1 Different Forms of Assignment

Mathematica has two operators for assigning values to symbols and expressions. The two forms are *lhs* = *rhs* and *lhs* := *rhs*. The first form, whose internal representation is Set[*lhs*, *rhs*], evaluates the right side before the assignment takes place. In the second form, internally represented as SetDelayed[*lhs*, *rhs*], the right side is assigned unevaluated. Evaluation takes place only later when the rule is used. This different behavior is achieved quite simply with the use of attributes.

Set[] does evaluate its second argument in the usual way. (More about the evaluation of its first argument in section 5.7.)

```
In[1]:= Attributes[Set]
Out[1]= {HoldFirst, Protected}
```

SetDelayed[] does not evaluate any of its arguments.

```
In[2]:= Attributes[SetDelayed]
Out[2]= {HoldAll, Protected}
```

Section 2.2.10 of the *Mathematica* book contains examples that explain when to use which of the two forms for assignment.

Delayed assignment is most often used when the left side contains patterns. This is the usual case with global transformation rules (see Chapter 6) and *Mathematica*'s equivalent for procedure or function definitions of traditional programming languages.

Assignments to symbols are usually not delayed. But there are two interesting uses of delayed assignment for symbols that we want to mention in the next two subsections.

■ 5.6.2 Application: Interface to a Window System

This subsection assumes familiarity with UNIX. Delayed assignments are used in the code for producing graphics on workstations with a window system. For example, **SunView.m** sets up *Mathematica* for producing its graphics under the SunView window system.

All graphics initialization files work by defining a value for $DisplayFunction, the default function used by Show[] to render its output. Typically, this value is the pure function Display[$Display, #]& and the device specific part is put in the definition for the file $Display. The function Display[*filename*, *graphics*] writes the POSTSCRIPT form of *graphics* to the file *filename*. If the first character of the file name is !, an external program is started that receives the graphics commands as its standard input. This program—sunps in this example—can have optional arguments. The three arguments are used to set the height and width of the window for the graphics and to set its title bar. We want to set the values of these arguments from inside *Mathematica*. This is done by assembling the file name under program control.

```
(* Graphics output for Sun under SunView *)

$DisplayWidth = 400
$DisplayHeight = 400
$DisplayTitle := StringForm["Mathematica: Out['1']", $Line]

$Display := $SunDisplay
$SunDisplay := StringJoin["!sunps -h ", ToString[$DisplayHeight],
                " -w ", ToString[$DisplayWidth],
                " -T '", ToString[$DisplayTitle], "'" ]

$DisplayFunction = Display[$Display, #]&
```

SunView.m

$Display is set to $SunDisplay using a delayed assignment and $SunDisplay in turn is set to the complicated StringJoin[] command shown above. Each time $Display is evaluated (which happens inside Show[]), the *current* values of the three global variables $DisplayWidth, $DisplayHeight, and $DisplayTitle are used to construct the argument string for sunps. $DisplayTitle, in turn, has a delayed value that uses the current value of $Line in StringForm[]. The effect of all this is that the line number of the command that produced the picture is put into the title bar. You should convince yourself that all delayed assignments are necessary. Note that no delayed assignment was necessary for assigning to $DisplayFunction itself, since $Display is used inside a pure function and is therefore not evaluated at this point.

Evaluating $Display gives the value that would be used inside Show[] to render graphics.	In[2]:= **$Display** Out[2]= !sunps -h 400 -w 400 -T 'Mathematica: Out[2]'
Evaluating it again gives a different value since $Line has changed.	In[3]:= **$Display** Out[3]= !sunps -h 400 -w 400 -T 'Mathematica: Out[3]'

■ 5.6.3 Application: A File of *Mathematica* Commands

Another use of delayed assignment for symbols is for assigning complicated commands to a number of symbols in a file. You can then read in that file and execute any of these commands by simply typing the corresponding symbol. The file **BookPictures.m** uses this. It has definitions for most of the pictures at the start of the chapters of this book. Its outline is as follows:

```
Needs["ComplexMap`"]
Needs["ParametricPlot3D`", "NewParametricPlot3D.m"]
⋮
chapter1 := PolarMap[ (2#-I)/(#-1)&, {0.001, 5.001, 0.25}, {0, 2Pi, Pi/15},
                 Framed->True ]
chapter2 :=
    ParametricPlot3D[{r*Cos[phi] - (r^2*Cos[2*phi])/2,
       -(r*Sin[phi]) - (r^2*Sin[2*phi])/2, (4*r^(3/2)*Cos[(3*phi)/2])/3},
       {r, 0.0001, 1, 0.9999/8}, {phi, 0, 4Pi, Pi/12}]
⋮
```

Part of **BookPictures.m**

First, all necessary auxiliary definitions are made and then a single assignment is made for each picture. An immediate assignment would compute all the graphics when you read in the file, which would take hours to do. Set up like this, each picture is only computed when you want to see it. The full text of **BookPictures.m** is reproduced in Appendix B.

In *Mathematica* versions with the *Notebooks* interface, there is a different way of achieving this. Simply put the commands into individual input cells that are not marked as initialization cells. To evaluate one of them, you select it and then choose the "evaluate selection" menu item. The auxiliary definitions at the beginning should be put into initialization cells.

■ 5.6.4 Multiple Assignments

It is possible to perform multiple assignments in one statement. Typically this looks like $sym_1 = sym_2 = expr$. This assigns the value of *expr* to both sym_1 and sym_2. Internally, it is represented as Set[sym_1, Set[sym_2, *expr*]]. The inner Set[] is evaluated first. Besides assigning the value of *expr* to sym_2 it also returns this value which is then assigned to sym_1.

With delayed assignments, it is quite a different case. Since the right side of a delayed assignment is not evaluated, the effect of sym_1 := sym_2 := *expr*, or in internal form SetDelayed[sym_1, SetDelayed[sym_2, *expr*]], is to assign the statement sym_2 := *expr* to sym_1. Nothing is assigned to sym_2 at this point. When sym_1 is evaluated later on the result of this evaluation would be to assign the unevaluated *expr* to sym_2. The value of the SetDelayed[] function itself is Null, since it cannot return anything meaningful. So the value of sym_1 is Null which is not printed.

The value of the outer assignment is `Null` which is not printed.	`In[1]:= a := b := e`

We have just defined this value for `a`.	`In[2]:= ?a`
	
	`a`
	`a/: a := b := e`

Using `a` performs the assignment to `b` and returns `Null`.	`In[2]:= a`
But now `b` too has a (delayed) value.	`In[3]:= b` `Out[3]= e`

The mixed form $sym_1 := sym_2 = expr$ is of some value when used for rules with patterns on the left side. See the section on dynamic programming (2.2.11) of the *Mathematica* book.

■ 5.7 Evaluation of the Left Side of an Assignment

In a rule of the form *lhs* -> *rhs* or *lhs* :> *rhs*, the left side *lhs* is evaluated in the normal way, like any other expression. In a definition of the form *lhs* = *rhs* or *lhs* := *rhs* things are different. In some applications, the exact way of evaluation of the left side is important. See also section B.4.2 of the Reference Guide.

■ 5.7.1 How the Left Side is Evaluated

In a definition like $f[e_1, e_2, \ldots, e_n] := rhs$, the left side $f[e_1, e_2, \ldots, e_n]$ is evaluated as follows. The arguments $e_1, e_2, \ldots, e_n$ of the top level function f are evaluated in the normal way. The top level function itself is *not* evaluated, i.e., no built-in code for f and no user-defined rules are applied. The definition is then made for this partially evaluated expression.

To see examples in which this is important, we need a way of comparing the expression before and after evaluation. To look at the internal form of an expression *before evaluation*, we can look at FullForm[Hold[*expr*]]. To look at the left side as it was used to define the rule, we look at the internal form of the rules attached to the head of the left side, f say, with FullForm[DownValue[f] /. (1_ :> _) :> 1]. (The substitution suppresses the right side of the rules in the value list.) Here is an example.

We want to define a rule with this left side, but first we look at the unevaluated form of it.	`In[1]:= FullForm[Hold[ f[x_ - y_] ]]`
	
	`Out[1]//FullForm=`
	`  Hold[f[Subtract[Pattern[x, Blank[]],`
	`      Pattern[y, Blank[]]]]]`

Now we define the rule.	`In[2]:= f[x_ - y_ ] := g[x, y]`

The subtraction in the left side has been replaced by addition and multiplication by -1.	`In[3]:= FullForm[DownValue[f] /. (1_ :> _) :> 1]`
	`Out[3]//FullForm=`
	`  ValueList[f[Plus[Pattern[x, Blank[]],`
	`      Times[-1, Pattern[y, Blank[]]]]]]`

This example also shows why the left side needs to be evaluated at all. Without evaluation the definition would be made for the pattern f[Subtract[x_, y_]]. Such a definition would never match, since in all expressions like f[a-b] the subtraction will have been turned into f[a + -1*b] before *Mathematica* tries to apply any definitions.

■ 5.7.2 Preventing the Evaluation

If the evaluation of an argument of the left side is undesirable, it can be suppressed by enclosing it in `Literal[]`. A definition used in subsection 7.2.3 has the form

> `N[Sum[args__], prec_] := NSum[args, AccuracyGoal -> prec]`.

In this form it would not work. The first argument in the left side is `Sum[args__]`. When this is evaluated it turns into `args__` and the definition would be made for `N[args__, prec_]` (which does not work). *Mathematica* functions do not do anything special if their arguments happen to be patterns. `Sum[args__]` is therefore treated like `Sum[e]` with *one* argument which evaluates to *e*. Therefore we have to use

> `N[Literal[Sum[args__]], prec_] := NSum[args, AccuracyGoal -> prec]`

which does not try to evaluate its first argument. The function `Literal[]` is removed before the definition is made and it does not interfere with the pattern.

`Sum[args__]` has been turned into `args__`.	`In[2]:= N[Sum[args__], prec_] :=`
	`        NSum[args, AccuracyGoal -> prec]`
	
	`N::nosym: N[args__, prec_]`
	`        does not contain a symbol to which to attach a rule.`

Now it works.	`In[3]:= N[Literal[Sum[args__]], prec_] :=` `        NSum[args, AccuracyGoal -> prec]`

`Literal[]` can also be used on the left side of a rule which is otherwise evaluated fully. To prevent any evaluation of the left side of a rule you can use

> *expr* `/. Literal[`*lhs*`] ->` *rhs* .

We have used this form in subsection 5.2.2, for example.

Chapter 6

Building Rule Sets

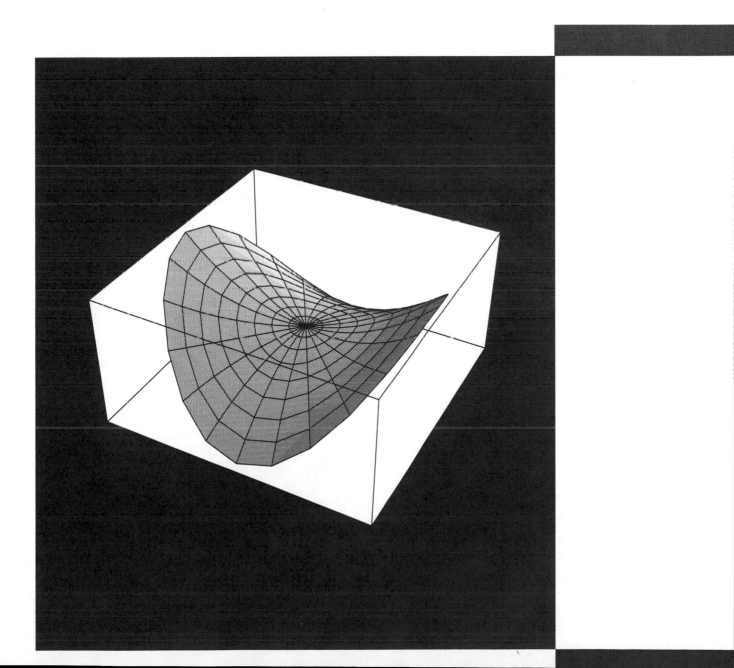

So far we have used definitions mostly in the way that other programming languages use procedures. Now we want to use rules to define simplifications and transformations of expressions. The ability to do this easily is probably the most important aspect of *Mathematica*.

Section 1 introduces the notions of *simplification* and *normal form*. Once we have identified a normal form for a class of expressions, we can give rules to transform other expressions into this normal form. The example we examine comes from trigonometry, which is especially rich in formulae.

Section 2 looks in more detail at how to write functions that apply a set of rules to an expression. Again trigonometry provides a good example.

In the next section, we look at how to define rules that will be automatically applied to all expressions.

Knowledge of *Mathematica*'s flexible pattern matching language is important for writing efficient rule sets. In section 4 we look at some advanced aspects of pattern matching for simplification rules and mathematical transformations.

Section 5 is about languages and grammars. In *Mathematica*, it is easy to define predicates that recognize certain classes of expressions. We explain the theory behind this and give an example.

About the illustration overleaf:

The saddle surface with a polar parameterization. In Cartesian coordinates the equation of the surface is $z = x^2 - y^2$. In cylindrical coordinates this becomes $z = r^2(\cos^2 \phi - \sin^2 \phi)$ which simplifies to $z = r^2 \cos(2\phi)$.

```
CylindricalPlot3D[r^2 Cos[2phi], {r, 0, 1/2, 1/16}, {phi, 0, 2Pi, 2Pi/24}]
```

■ 6.1 Introduction to Simplification Rules and Normal Forms

A good source of examples for simplification rules is *trigonometry*. There are a lot of identities between expressions involving the trigonometric functions Sin[], Cos[], and Tan[].

■ 6.1.1 The Normal Form of Expressions

An important concept in simplifying expressions is that of a *normal form* of an expression. The number 0, for example, could be written in many different ways, as 0, $1 - 1$, $i^2 + 1$, or $\cos(\pi/2)$. All are equivalent and we have a clear idea about which is the simplest. Further, we know how to transform all others into the simplest form.

If we can define a normal form for a class of expressions, then two different but equivalent expressions can both be simplified to this normal form and it is therefore easy to decide whether two expressions are equivalent.

A normal form need not be the intuitively simplest form. For polynomials, the fully expanded form is a normal form. We know how to expand polynomials and if two polynomials are equal, then their expanded form is the same. Yet for many polynomials, we would consider a factored form to be simpler.

To decide whether these two polynomials are the same we expand them and compare the results.

```
In[1]:= {x^2-1, (x-1)(x+1)}

Out[1]= {-1 + x , (-1 + x) (1 + x)}
                2
```

They are indeed equal.

```
In[2]:= Expand[%]

Out[2]= {-1 + x , -1 + x }
                2        2
```

The normal—expanded—form of $(x + 1)^5$ is more "complicated" than the factored form.

```
In[3]:= Expand[(x+1)^5]

Out[3]= 1 + 5 x + 10 x  + 10 x  + 5 x  + x
                      2        3       4     5
```

Here the expanded form is simpler.

```
In[4]:= Expand[(-1 + x)(1 + x)(1 + x^2)(1 + x^4)]

Out[4]= -1 + x
               8
```

If there are several ways of writing an expression, we should try to find one that can serve as the normal form. A necessary condition is that we have rules for reducing all other forms to this form and that the rules applied to the normal form itself do not change it. Otherwise, we would go on and on applying rules without an end.

We know that Sin[-*x*] is the same as -Sin[*x*]. We can therefore decide that the normal form for trigonometric functions should have nonnegative arguments. Here are the rules that perform these simplifications:

```
TrigCanonicalRules = {
    Sin[x_?Negative] :> - Sin[-x],
    Cos[x_?Negative] :>   Cos[-x],
    Tan[x_?Negative] :> - Tan[-x]
}
```

Simplification of trigonometric functions of negative arguments

With these rules we can perform a few trivial simplifications.

Here is a trigonometric expression.

```
In[2]:= Sin[2] - Sin[-3] + 2 Cos[-1/2] - Cos[1/2]

              1            1
Out[2]= 2 Cos[-(-)] - Cos[-] - Sin[-3] + Sin[2]
              2            2
```

Applying the rules puts it in standard form. The usual arithmetic rules then perform further simplifications, in this case turning 2 Cos[-1/2] - Cos[1/2] into Cos[1/2].

```
In[3]:= % /. TrigCanonicalRules

             1
Out[3]= Cos[-] + Sin[2] + Sin[3]
             2
```

No simplification is performed here since the argument -a does not satisfy the predicate Negative[]. More rules are necessary for dealing with symbolic arguments.

```
In[4]:= Sin[-a] /. TrigCanonicalRules

Out[4]= Sin[-a]
```

For symbolic arguments we cannot tell whether they are negative or not. We therefore decide that the normal form is one where there is no *explicit* minus sign in the argument.

This is our first try at a rule for symbolic expressions. It works here.

```
In[5]:= Sin[-a] /. Sin[-x_] :> - Sin[x]
Out[5]= -Sin[a]
```

But it is not general enough.

```
In[6]:= Sin[-2a] /. Sin[-x_] :> - Sin[x]
Out[6]= Sin[-2 a]
```

The full form of the expression Sin[-2a] and the pattern in the rule show why the pattern did not match.

```
In[7]:= FullForm[{Sin[-2a], Sin[-x_]}]
Out[7]//FullForm=
  List[Sin[Times[-2, a]],
    Sin[Times[-1, Pattern[x, Blank[]]]]]
```

This pattern is more general. It allows for any negative number in the product.

```
In[8]:= Sin[-2a] /. Sin[n_?Negative x_] :> - Sin[-n x]
Out[8]= -Sin[2 a]
```

By making x_ optional, this rule can even deal with the old case of a purely numerical argument.

```
In[9]:= Sin[-2] /. Sin[n_?Negative x_.] :> - Sin[-n x]

Out[9]= -Sin[2]
```

Here are the improved rules that deal with numerical and symbolic arguments:

```
TrigCanonicalRules = {
    Sin[n_?Negative x_.] :> -Sin[-n x],
    Cos[n_?Negative x_.] :>  Cos[-n x],
    Tan[n_?Negative x_.] :> -Tan[-n x]
}
```

Simplifications for negative and symbolic arguments

■ 6.1.2 Ordering of Expressions

If you type in the expression b+a, then *Mathematica* turns it into a+b. This form can hardly be considered simpler but the built-in ordering function nevertheless provides a normal form for sums and products. This concept is quite powerful indeed. By sorting all the terms of a long sum into standard order, it is quite easy to combine terms that are the same and also perform all the numerical additions that are possible, since numbers are sorted first.

Continuing our example with trigonometric functions, we can now define a normal form for trigonometric functions of arguments that are sums. The normal form shall not have a minus sign in the *first* term of the sum.

This rule works in this simple case.

```
In[1]:= Sin[-a + b] /. Sin[n_?Negative x_ + y_] :>
                 - Sin[-n x - y]

Out[1]= Sin[a   b]
```

But it also transforms this expression which according to our definition is already in normal form (since the first term has no minus sign).

```
In[2]:= Sin[a - b] /. Sin[n_?Negative x_ + y_] :>
                 - Sin[-n x - y]

Out[2]= -Sin[-a + b]
```

To avoid this, we use the ordering predicate as a side condition for the rule.

```
In[3]:= Sin[a - b] /. Sin[n_?Negative x_ + y_] :>
                 - Sin[-n x - y] /; OrderedQ[{x, y}]

Out[3]= Sin[a - b]
```

It still simplifies this expression as it should, since the first term is -a.

```
In[4]:= Sin[b - a] /. Sin[n_?Negative x_ + y_] :>
                 - Sin[-n x - y] /; OrderedQ[{x, y}]

Out[4]= -Sin[a - b]
```

This time however we cannot easily make `x_` optional to deal with numbers. If we did, this expression would not have a normal form.

```
In[5]:= Sin[1 - a - b] /. Sin[n_?Negative x_. + y_] :>
                            - Sin[-n x - y] /; OrderedQ[{x, y}]

Out[5]= -Sin[-1 + a + b]
```

Applying the rule again turns the expression back into the original. Using such a rule would lead to an infinite loop.

```
In[6]:= % /. Sin[n_?Negative x_. + y_] :>
                            - Sin[-n x - y] /; OrderedQ[{x, y}]

Out[6]= Sin[1 - a - b]
```

Here is our new rule set:

```
TrigCanonicalRules = {
    Sin[n_?Negative x_.] :> -Sin[-n x],
    Cos[n_?Negative x_.] :>  Cos[-n x],
    Tan[n_?Negative x_.] :> -Tan[-n x],
    Sin[n_?Negative x_ + y_] :> - Sin[-n x - y] /; OrderedQ[{x, y}],
    Cos[n_?Negative x_ + y_] :>   Cos[-n x - y] /; OrderedQ[{x, y}],
    Tan[n_?Negative x_ + y_] :> - Tan[-n x - y] /; OrderedQ[{x, y}]
}
```

Normal forms for sums as arguments

In all the following rule sets we will always apply these `TrigCanonicalRules` to all the expressions to bring them into normal form.

■ 6.2 Trigonometric Simplifications

■ 6.2.1 Expansion of Products and Powers

In this section we will use identities like

$$\sin x \sin y = \frac{\cos(x - y)}{2} - \frac{\cos(x + y)}{2}$$

to write expressions involving products and powers of trigonometric functions in a form in which all the trigonometric functions occur only linearly.

Here are the rules we need to write products of `Sin[]` and `Cos[]` as sums:

```
TrigLinearRules = {
    Sin[x_] Sin[y_] :> Cos[x-y]/2 - Cos[x+y]/2,
    Cos[x_] Cos[y_] :> Cos[x+y]/2 + Cos[x-y]/2,
    Sin[x_] Cos[y_] :> Sin[x+y]/2 + Sin[x-y]/2
}
```

Writing products of trigonometric functions as sums

All the products are rewritten as sums of trigonometric functions.

```
In[2]:= Sin[a] Cos[b] + Sin[a] Cos[a] + Cos[2a] Cos[3a] /.
            TrigLinearRules
```

$$Out[2]= \frac{Cos[-a]}{2} + \frac{Cos[5\ a]}{2} + \frac{Sin[2\ a]}{2} + \frac{Sin[a\ -\ b]}{2} +$$

$$\frac{Sin[a\ +\ b]}{2}$$

The operator `/.` applies the rules only once to a particular subexpression.

```
In[3]:= Sin[a] Sin[2a] Sin[3a] Sin[4a] /. TrigLinearRules
```

$$Out[3]= (\frac{Cos[-a]}{2} - \frac{Cos[3\ a]}{2})\ Sin[3\ a]\ Sin[4\ a]$$

`//.` applies them over and over again. The result is not yet in final form, since there are still products of trigonometric functions around.

```
In[4]:= Sin[a] Sin[2a] Sin[3a] Sin[4a] //. TrigLinearRules
```

$$Out[4]= (\frac{Cos[-a]}{2} - \frac{Cos[3\ a]}{2})\ (\frac{Cos[-a]}{2} - \frac{Cos[7\ a]}{2})$$

We want to repeatedly apply the rules and expand the result until it no longer changes. `FixedPoint[]` makes this easy.

```
In[5]:= FixedPoint[ Expand[# //. TrigLinearRules]&,
            Sin[a] Sin[2a] Sin[3a] Sin[4a] ]
```

$$Out[5]= \frac{-Cos[-8\ a]}{8} + \frac{Cos[-a]^2}{4} - \frac{Cos[2\ a]}{8} - \frac{Cos[6\ a]}{8} +$$

$$\frac{Cos[10\ a]}{8}$$

The last example above shows that we need special rules for powers of trigonometric functions. Mathematically Sin[*x*]∧2 is, of course, equal to Sin[*x*] Sin[*x*], but its representation is different and so our rules do not match. We can derive the necessary rules for Sin[*x*]∧*n* and Cos[*x*]∧*n* from the rules we already have. We treat Sin[*x*]∧*n* as Sin[*x*] Sin[*x*] Sin[*x*]∧(*n*-2) and then apply the rule for Sin[*x*] Sin[*y*]. The new rules are:

```
Sin[x_]^n_Integer?Positive :> (1/2 - Cos[2x]/2) Sin[x]^(n-2)
Cos[x_]^n_Integer?Positive :> (1/2 + Cos[2x]/2) Cos[x]^(n-2)
```

Powers of trigonometric functions

The restriction n_Integer?Positive for the exponent is necessary since negative exponents or rational powers would lead to infinite loops.

Since it is now necessary to repeatedly apply the rules and expand the intermediate results, it makes sense to write a function that does these steps for us. We also want to apply the rules that put the result into normal form. In addition, we put everything into a package.

```
BeginPackage["TrigSimplification`"]

TrigNormal::usage = "TrigNormal[e] puts expressions with trigonometric
    functions into normal form."

TrigLinear::usage = "TrigLinear[e] expands products and powers of trigonometric
    functions."

Begin["`Private`"]

TrigCanonicalRules = {
    Sin[n_?Negative x_.] :> -Sin[-n x],
    Cos[n_?Negative x_.] :>  Cos[-n x],
    Tan[n_?Negative x_.] :> -Tan[-n x],
    Sin[n_?Negative x_ + y_] :> - Sin[-n x - y] /; OrderedQ[{x, y}],
    Cos[n_?Negative x_ + y_] :>   Cos[-n x - y] /; OrderedQ[{x, y}],
    Tan[n_?Negative x_ + y_] :> - Tan[-n x - y] /; OrderedQ[{x, y}]
}

TrigLinearRules = {
    Sin[x_] Sin[y_] :> Cos[x-y]/2 - Cos[x+y]/2,
    Cos[x_] Cos[y_] :> Cos[x+y]/2 + Cos[x-y]/2,
    Sin[x_] Cos[y_] :> Sin[x+y]/2 + Sin[x-y]/2,
    Sin[x_]^n_Integer?Positive :> (1/2 - Cos[2x]/2) Sin[x]^(n-2),
    Cos[x_]^n_Integer?Positive :> (1/2 + Cos[2x]/2) Cos[x]^(n-2)
}

SetAttributes[{TrigNormal, TrigLinear}, Listable]
```

```
TrigNormal[e_] := e /. TrigCanonicalRules
TrigLinear[e_] :=
    FixedPoint[ Expand[# //. TrigLinearRules /. TrigCanonicalRules]&, e ]
End[]
Protect[TrigNormal, TrigLinear]
EndPackage[]
```

The first version of **TrigSimplification1.m**

This performs many simplification steps to get to the final result.

In[2]:= **TrigLinear[Sin[x]^2 Cos[x]^3]**

$$Out[2]= \frac{Cos[x]}{8} - \frac{Cos[3\ x]}{16} - \frac{Cos[5\ x]}{16}$$

Integrating and then differentiating a trigonometric expression should give the old expression back.

In[3]:= **Integrate[Sin[x]^2 Cos[x]^2, x]**

$$Out[3]= \frac{x}{8} - \frac{Cos[x]\ Sin[x]}{8} + \frac{Cos[x]\ Sin[x]^3}{4}$$

It looks quite different, however.

In[4]:= **D[%, x]**

$$Out[4]= \frac{1}{8} - \frac{Cos[x]^2}{8} + \frac{Sin[x]^2}{8} + \frac{3\ Cos[x]^2\ Sin[x]^2}{4} - \frac{Sin[x]^4}{4}$$

We put the result into normal form.

In[5]:= **TrigLinear[%]**

$$Out[5]= \frac{1}{8} - \frac{Cos[4\ x]}{8}$$

We also put the argument of the integral into normal form. The two are indeed the same.

In[6]:= **TrigLinear[Sin[x]^2 Cos[x]^2]**

$$Out[6]= \frac{1}{8} - \frac{Cos[4\ x]}{8}$$

The last example shows an important technique for proving that two expressions are the same. Instead of trying to convert one into the other which can be very difficult or even impossible we transform both to a common normal form.

■ 6.2.2 Simplifying the Arguments of Trigonometric Functions

`TrigLinear[]` linearizes trigonometric expressions and in doing so may introduce more complicated arguments of the trigonometric functions.

The simple arguments x and y of `Sin[]` and `Cos[]` are turned into more complicated ones.	`In[2]:= TrigLinear[ Cos[x] Cos[y] Sin[x] ]`

$$Out[2]= -\frac{Sin[2\ x - y]}{4} + \frac{Sin[2\ x + y]}{4}$$

We might want to simplify the arguments instead of linearizing the expression. This is possible through the use of the two formulae

$$\sin(x + y) = \sin x \cos y + \cos x \sin y$$
$$\sin(x + y) = \cos x \cos y - \sin x \sin y.$$

It is easy to write down two rules that implement these transformations. Again we have to be careful about special cases. $\sin(2x)$ is the same as $\sin(x + x)$ and so we can use these formulae but we need a different pattern to match this case. More generally, we treat $\sin nx$ as $\sin(x + (n - 1)x)$ for positive integer n and let repeated rule application unwind this case. Note that we do not need to give rules for negative multiples of x, since we put the trigonometric functions into normal form first and the normal forms do not have negative arguments. Here is the rule set and the function that applies them:

```
TrigArgument::usage = "TrigArgument[e] writes trigonometric functions of multiple
    angles as products of functions of that angle."
:

TrigArgumentRules = {
    Sin[x_ + y_] :> Sin[x] Cos[y] + Sin[y] Cos[x],
    Cos[x_ + y_] :> Cos[x] Cos[y] - Sin[x] Sin[y],
    Sin[n_Integer?Positive x_.] :> Sin[x] Cos[(n-1)x] + Sin[(n-1)x] Cos[x],
    Cos[n_Integer?Positive x_.] :> Cos[x] Cos[(n-1)x] - Sin[x] Sin[(n-1)x]
}
:

TrigArgument[e_] :=
    Together[ FixedPoint[ (# //. TrigArgumentRules /. TrigCanonicalRules)&, e ] ]
```

Part of **TrigSimplification2.m**: Reducing the complexity of arguments

In the function `TrigArgument[]` that applies the rules, we use `Together[]` to beautify the result.

We can get the expression we started with in input line 2 by undoing the effect of `Triglinear[]`.

```
In[3]:= TrigArgument[%]
Out[3]= Cos[x] Cos[y] Sin[x]
```

Here is another example. First we expand the expression.

```
In[4]:= TrigLinear[Sin[x]^2]
Out[4]= 1   Cos[2 x]
        - - --------
        2      2
```

And then we try to get back to the original one. The result, however, looks different.

```
In[5]:= TrigArgument[%]
                   2        2
         1 - Cos[x]  + Sin[x]
Out[5]= ----------------------
                  2
```

To prove that it is correct we put it into normal form.

```
In[6]:= TrigLinear[%]
Out[6]= 1   Cos[2 x]
        - - --------
        2      2
```

The last example shows that `TrigArgument[]` does *not* produce normal forms! There are many different ways to write a trigonometric expression if we allow products of trigonometric functions to occur. This is a consequence of identities like $\sin^2 x + \cos^2 x = 1$.

■ 6.2.3 Performance Considerations

In our rules for `TrigLinear[]` and `TrigArgument[]` we have expressed `Sin[x]^n` in terms of `Sin[x]^(n-2)` and `Sin[n x]` in terms of `Sin[(n-1) x]`. This has the advantage of being very easy to write. We then let repeated rule application take care of simplifying expressions with large `n`. It is, however, rather inefficient as timing experiments show.

This generates a table of the time it takes to apply our rules to the expressions $\sin(nx), n = 1, 2, \ldots, 10$. The time roughly doubles for each successive n.

```
In[2]:= Table[ Timing[ TrigArgument[Sin[n x]] ][[1]],
               {n, 10} ]
Out[2]= {0.05 Second, 0.1 Second, 0.3 Second,
         0.683333 Second, 1.46667 Second, 3.06667 Second,
         6.31667 Second, 12.75 Second, 25.6333 Second,
         51.6833 Second}
```

We can do better by applying a formula for $\sin(nx)$ and $\cos(nx)$ that will directly generate an expression involving only $\sin x$ and $\cos x$. Any handbook of mathematics will contain such a formula.

$$\sin(nx) \;=\; n\cos^{n-1}x - \binom{n}{3}\cos^{n-3}x\,\sin^3 x + \binom{n}{5}\cos^{n-5}x\,\sin^5 x - \cdots$$

$$\cos(nx) \;=\; \cos^n x - \binom{n}{2}\cos^{n-2}x\,\sin^2 x + \binom{n}{4}\cos^{n-4}x\,\sin^4 x - \cdots .$$

An important design goal of *Mathematica* was to make it easy to program such formulae as rules. Here is a straightforward translation:

```
Sin[n_Integer?Positive x_.] :>
    Sum[ (-1)^((i-1)/2) Binomial[n, i] Cos[(n-i) x] Sin[i x], {i, 1, n, 2} ]
Cos[n_Integer?Positive x_.] :>
    Sum[ (-1)^(i/2) Binomial[n, i] Cos[(n-i) x] Sin[i x], {i, 0, n, 2} ]
```

A mathematical formula translated into *Mathematica*

TrigSimplification3.m contains these rules instead of the old ones.

The time it takes is now almost linear in n.

```
In[2]:= Table[ Timing[ TrigArgument[Sin[n x]] ][[1]],
            {n, 10} ]

Out[2]= {0.0666667 Second, 0.116667 Second,
    0.216667 Second, 0.216667 Second, 0.266667 Second,
    0.316667 Second, 0.383333 Second, 0.416667 Second,
    0.466667 Second, 0.5 Second}
```

If such a formula were not available, we could still improve the performance. Instead of treating $\sin(nx)$ as $\sin((n-1)x)\sin x$, we could break it apart in the middle, provided that n is even, and write it as $\sin(\frac{n}{2}x + \frac{n}{2}x)$. If n is odd, we split it as $\frac{n+1}{2}$ and $\frac{n-1}{2}$. Here are the rules; they are part of **TrigSimplification4.m**:

```
    Sin[n_Integer?EvenQ x_.] :>
        Sin[n/2 x] Cos[n/2 x] + Sin[n/2 x] Cos[n/2 x],
    Sin[n_Integer?OddQ x_.] :>
        Sin[(n+1)/2 x] Cos[(n-1)/2 x] + Sin[(n-1)/2 x] Cos[(n+1)/2 x],
    Cos[n_Integer?EvenQ x_.] :>
        Cos[n/2 x] Cos[n/2 x] - Sin[n/2 x] Sin[n/2 x],
    Cos[n_Integer?OddQ x_.] :>
        Cos[(n+1)/2 x] Cos[(n-1)/2 x] - Sin[(n+1)/2 x] Sin[(n-1)/2 x]
```

Another way of treating `Sin[n x]`

Note that we dropped the condition that n be positive, since these rules happen to work also for negative n. We would not need them for negative n because the canonicalization turns all negative arguments into positive ones (see section 6.1). They do not lead to infinite loops in this case as the old set of rules would.

We notice that the timings are quite irregular, depending on the binary representation of n. The theoretical complexity is $n \log n$, somewhere between the two previous methods.

```
In[2]:= Table[ Timing[ TrigArgument[Sin[n x]] ][[1]],
               {n, 10} ]
Out[2]= {0.0666667 Second, 0.133333 Second,
    0.383333 Second, 0.3 Second, 1.11667 Second,
    0.816667 Second, 1.5 Second, 0.716667 Second,
    2.96667 Second, 2.28333 Second}
```

The standard package **Trigonometry.m** that comes with *Mathematica* contains rule sets like the ones we have looked at in this section to transform trigonometric expressions in many different ways.

■ 6.3 Globally Defined Rules

In the preceding section, we have given an example of a rule set, a collection of rules and functions to apply them to expressions. Thus *Mathematica* will not use these rules on its own, rather you have to give a command to apply them to an expression. We have in effect defined a new function that just happens to be implemented using rules instead of some of the traditional programming constructs.

Now we want to look at another approach. We will set things up so that *Mathematica* automatically tries to use a set of rules on all expressions it evaluates. The advantage of this approach is that you do not have to explicitly call a function to get things done. On the downside, it is almost impossible to prevent such global rules from being applied to a certain expression should you want to do that.

■ 6.3.1 Example: Trigonometric Simplification Again

Let us go back to the problem of putting trigonometric expressions in normal form. A set of rules for doing this was given at the end of subsection 6.1.2 on page 138. It is rather straightforward to turn these rules into global definitions.

```
Unprotect[Sin, Cos, Tan]

Sin[n_?Negative x_.] := -Sin[-n x]
Cos[n_?Negative x_.] :=  Cos[-n x]
Tan[n_?Negative x_.] := -Tan[-n x]
Sin[n_?Negative x_ + y_] := - Sin[-n x - y] /; OrderedQ[{x, y}]
Cos[n_?Negative x_ + y_] :=   Cos[-n x - y] /; OrderedQ[{x, y}]
Tan[n_?Negative x_ + y_] := - Tan[-n x - y] /; OrderedQ[{x, y}]

Protect[Sin, Cos, Tan]
```

TrigDefine1.m: Definitions for normal forms of trigonometric functions

Note that we have to unprotect the trigonometric functions before we can define any rules for them. After defining these rules, any trigonometric function encountered will be put into normal form.

All arguments of trigonometric functions are made nonnegative.

```
In[2]:= Sin[-x] Cos[y-x] + 1/Cos[-2]

             1
Out[2]=  ───────  - Cos[x - y] Sin[x]
           Cos[2]
```

Now we turn this rule set into a proper package. Even though we do not export any functions from this package, we still define a context for it. The reason is that only in this case can we read it in using Needs["*Context*`"]. We can follow the suggestions given in the skeletal package in section 2.4.

```
BeginPackage["TrigDefine`"]

TrigDefine::usage = "TrigDefine.m defines global rules for putting
    arguments of trigonometric functions into normal form."

Begin["`Private`"]    (* begin the private context *)

(* unprotect any system functions for which rules will be defined *)

protected = Unprotect[Sin, Cos, Tan]

Sin[n_?Negative x_.] := -Sin[-n x]
Cos[n_?Negative x_.] :=  Cos[-n x]
Tan[n_?Negative x_.] := -Tan[-n x]
Sin[n_?Negative x_ + y_] := - Sin[-n x - y] /; OrderedQ[{x, y}]
Cos[n_?Negative x_ + y_] :=   Cos[-n x - y] /; OrderedQ[{x, y}]
Tan[n_?Negative x_ + y_] := - Tan[-n x - y] /; OrderedQ[{x, y}]

Protect[ Release[protected] ]      (* restore protection of system symbols *)

End[]         (* end the private context *)

EndPackage[]  (* end the package context *)
```

TrigDefine2.m: A package for trigonometric definitions

We can also turn the rules for expanding products and powers of trigonometric functions from section 6.2 into global rules.

```
Sin/: Sin[x_] Sin[y_] := Cos[x-y]/2 - Cos[x+y]/2
Cos/: Cos[x_] Cos[y_] := Cos[x+y]/2 + Cos[x-y]/2
Sin/: Sin[x_] Cos[y_] := Sin[x+y]/2 + Sin[x-y]/2

Sin/: Sin[x_]^n_Integer?Positive := Expand[(1/2 - Cos[2x]/2) Sin[x]^(n-2)]
Cos/: Cos[x_]^n_Integer?Positive := Expand[(1/2 + Cos[2x]/2) Cos[x]^(n-2)]
```

Part of **TrigDefine3.m**

If you compare these definitions with the corresponding rules in **TrigSimplification1.m** on page 141 you will notice the extra Expand[] on the right side of the definitions for powers of Sin[] and Cos[]. This ensures that higher powers will be simplified fully.

Note the declarations Sin/: and Cos/: on the left side of the definitions. Without them definitions would be stored with the top level operation of the left side. For the rules above this would be Times[] or Power[]. It is normally not a good idea to store rules with the basic arithmetic operations. This slows down *every* multiplication and exponentiation performed. Rather they should be stored with the operands to which they apply.

Reading in the file sets up our rules.   In[1]:= << TrigDefine3.m

All possible rules are applied.   In[2]:= Sin[a] Cos[b]

$$Out[2]= \frac{Sin[a - b]}{2} + \frac{Sin[a + b]}{2}$$

Powers are expanded fully, but the result looks ugly.   In[3]:= Sin[alpha]^4

$$Out[3]= \frac{1}{4} - \frac{Cos[2\ alpha]}{2} + \frac{\frac{1}{2} + \frac{Cos[4\ alpha]}{2}}{4}$$

Another expansion is necessary to get the simplest form. In[4]:= Expand[%]

$$Out[4]= \frac{3}{8} - \frac{Cos[2\ alpha]}{2} + \frac{Cos[4\ alpha]}{8}$$

This product is not expanded fully. There are still products of trigonometric functions around. However, none of the rules given matches.   In[5]:= Cos[alpha] Cos[beta] Sin[beta]

$$Out[5]= (\frac{Cos[alpha - beta]}{2} + \frac{Cos[alpha + beta]}{2})\ Sin[beta]$$

After expanding the products the rules once more match and fully linearize the expression. It is still in a rather complicated form, however, needing another Expand[].   In[6]:= Expand[%]

$$Out[6]= \frac{\frac{Sin[alpha]}{2} - \frac{Sin[alpha - 2\ beta]}{2}}{2} +$$

$$\frac{\frac{-Sin[alpha]}{2} + \frac{Sin[alpha + 2\ beta]}{2}}{2}$$

We can define a function SuperExpand[] as the fixed point of Expand[] that keeps expanding terms until they no longer change.   In[7]:= SuperExpand = FixedPoint[Expand, #]&

Out[7]= FixedPoint[Expand, #1] &

This is the form we want.   In[8]:= SuperExpand[Cos[alpha] Cos[beta] Sin[beta]]

$$Out[8]= \frac{-Sin[alpha - 2\ beta]}{4} + \frac{Sin[alpha + 2\ beta]}{4}$$

The function TrigLinear[] gave us greater control over the evaluation process. There we just kept applying the rules and expanding until the expression no longer changed. This is not so easy to achieve with global definitions.

■ 6.4 Pattern Matching for Rules

In Chapter 1 and Chapter 3 we have already used some of the possibilities of pattern matching. There we defined rules for procedures, for example `CartesianMap[]` or `PolarMap[]`. These rules look like $f[arg_1, arg_2, \ldots] := body$. In principle, every rule is of this form but now we take a different point of view. We do not view a rule like `Sin[x_] Cos[y_] := ...` encountered in subsection 6.3.1 as defining a procedure for `Times[]`, the top-level operation of the left side. Rather, we view it as specifying an arbitrarily complicated pattern that is to be replaced by the right side of the rule whenever it occurs.

For technical reasons, each rule has to be associated with a symbol. A definition like `f[x_] := ` *body* naturally belongs to `f`, the head of the left side. If the head of the left side is an arithmetic operation, as in `Sin[x_] Cos[y_] := ...` then the rule should be associated with one of the arguments, if possible. The symbol with which the rule is to be associated must either be the head of the left side (the default) or the head of one of the arguments of the top-level operation.

■ 6.4.1 Patterns for Mathematical Formulae

When we give a formula like

$$\sin(x + y) = \sin x \cos y + \cos x \sin y$$

then we are quite good at recognizing the expression $\sin(x + y)$ even when it comes in disguise as in $\sin(2\alpha)$ or $\sin(a + b + c)$ and can apply the formula even in these cases. The corresponding definition in *Mathematica* is

$$\text{Sin[x\_ + y\_] := Sin[x] Cos[y] + Sin[y] Cos[x]}$$

as we have seen in subsection 6.2.2. In internal form, the left side of this rule is `Sin[Plus[x_, y_]]`, while the two examples given are `Sin[Times[2, alpha]]` and `Sin[Plus[a, b, c]]`, and so the rule would not match since there is no way of filling in expressions for `x_` and `y_` that make `Sin[Plus[x_, y_]]` equal to either of the examples. While we do in fact need a special rule to match cases like $\sin(2\alpha)$, we do not need special rules for the second case. In the following two subsections, we look at the two facilities that *Mathematica* provides for making the design of rules easier: attributes and defaults.

■ 6.4.2 Pattern Matching for Flat and Orderless Functions

The two arithmetic operations addition and multiplication have the attributes `Flat` and `Orderless` defined. This causes their arguments to be sorted in standard order and nested expressions to be flattened out. For example, the expression `(b + c) + a` or `Plus[Plus[b, c], a]` is first turned into `Plus[b, c, a]` and then sorted to give `Plus[a, b, c]`. Pattern matching takes these attributes into account and can reverse this process. To match `Sin[Plus[a, b, c]]` with `Sin[Plus[x_, y_]]`, it is first turned into `Sin[Plus[a, Plus[b, c]]]` and then it matches with `x_` becoming a and `y_` becoming b+c.

This rule does not do anything particularly useful, but it shows us how the pattern matches by returning the values of the two slots `x_` and `y_`.	`In[1]:= f[x_ + y_] := {x, y}`
This is as expected.	`In[2]:= f[a + b]` `Out[2]= {a, b}`
Mathematica applies the attribute `Flat` in reverse to find a match. There is no guarantee as to how this will be done, as `(a+b)+c` or `a+(b+c)`.	`In[3]:= f[a + b + c]` `Out[3]= {a, b + c}`

More examples of this can be found in section 2.3.1 of the *Mathematica* book.

■ 6.4.3 Defaults for Arithmetic Operations

We have seen that we can view `Sin[a + b + c]` as an instance of `Sin[x_ + y_]`. But what about `Sin[a]`? *Mathematica* does not treat this as `Sin[a + 0]` and so the rule would not match. In fact, we would not want it to match since the right side would be `Sin[a] Cos[0] + Sin[0] Cos[a]` that simplifies back to `Sin[a]`.

In other cases, this would be quite useful. For example, we can define a rule for finding the derivatives of powers as

$$\texttt{diff[x\_\^n\_, x\_] := n x\^(n-1)}.$$

This rule would not work for the case `diff[x, x]`, since the first argument is not a power. We know how to make it work in this case: we can declare the exponent optional by using `x_^n_.` instead of `x_^n_` as above.

This is almost the same definition as in the preceding subsection except that now the second term in the sum is optional.	`In[1]:= f[x_ + y_.] := {x, y}`
This is treated as `f[a + 0]` and the rule matches even though `Plus[]` does not occur in the expression at all!	`In[2]:= f[a]` `Out[2]= {a, 0}`

Defaults are defined for addition, multiplication, and exponentiation. You can define defaults for your own functions.

The default for arguments of p is now set to 17.	`In[3]:= Default[p] = 17`
	
	`Out[3]= 17`

Such an assignment is automatically stored with p and not with `Default`.	`In[4]:= ?p`
	
	`p`
	`p/: Default[p] = 17`

Here is a rule involving default arguments for p.	`In[4]:= g[p[x_, y_.]] := {x, y}`

The default is supplied and the rule matches.	`In[5]:= g[p[5]]`
	
	`Out[5]= {5, 17}`

p must be present however for the rule to match.	`In[6]:= g[a]`
	
	`Out[6]= g[a]`

To achieve the same effect as with `Plus[]` above, we need to define the attribute `OneIdentity` for our function (addition and multiplication both have this attribute set).

We will define the same rules for q as we did for p but in addition we also define the attribute `OneIdentity`.	`In[7]:= SetAttributes[q, OneIdentity]`

The default for arguments of q is now set to 18.	`In[8]:= Default[q] = 18`
	
	`Out[8]= 18`
	
	`In[9]:= g[q[x_, y_.]] := {x, y}`

It matches even though q is not there at all.	`In[10]:= g[a]`
	
	`Out[10]= {a, 18}`

`Plus[]` has all these attributes and defaults already defined.	`In[11]:= ??Plus`
	
	`x + y + z represents a sum of terms.`
	`Attributes[Plus] =`
	
	`  {Flat, Listable, OneIdentity, Orderless, Protected}`
	`Plus/: Default[Plus] := 0`

`OneIdentity` means that $f[x]$ is the same as x. The expression `g[a]` can therefore be turned into `g[q[a]]`. Then the default value for the second argument of q is used to match the rule.

■ 6.4.4 Conditional Pattern Matching

There are two ways of restricting the expressions that match a pattern. You can give a predicate in the form $f[x\_?pred]$ directly in the pattern on the left side of a rule or you can put a condition on the right side of a rule in the form $f[x\_]$:= *expr* /; *condition*. The former is used to restrict the matching of a single slot in the pattern. The second form is used for more complicated cases where the condition involves the values of several variables from the left side.

Another important case is the restriction of the matches to a certain *type* of expressions, identified by their head. To make a function accept only integer arguments, you can use f[n_Integer] := *body*. This can be combined with a predicate, for example, restricting the argument to a positive integer with

$$f[n\_Integer?Positive] := body.$$

Apart from the better performance this is also good programming style, putting all requirements for the arguments of the function in one place, and is preferable to

$$f[n\_Integer] := body \text{ /; Positive[n]}.$$

If no built-in predicate exists, you can either define one or use a pure function. In this case, it might be better to use a condition. To accept only integers greater than 3, we would write f[n_Integer?(#>3&)] := *body* or f[n_Integer] := *body* /; n > 3. Note that the whole pure function has to be put in parentheses since the priority of ? is higher than the priority of &.

■ 6.4.5 Subtraction and Division

All subtractions are evaluated to an addition and multiplication by -1. The expression $a - b$ or Subtract[a, b] evaluates to $a + -1\ b$. This normally *prints* again as $a - b$ and therefore this evaluation is not quite obvious. A division of the form a/b is transformed into $a\ b\wedge-1$ already by the parser.

This is the form of a subtraction *before* evaluation.	In[1]:= **FullForm[Hold[a - b]]** Out[1]//FullForm= Hold[Subtract[a, b]]
It evaluates to this expression.	In[2]:= **FullForm[a - b]** Out[2]//FullForm= Plus[a, Times[-1, b]]
A division is transformed into this form already on input.	In[3]:= **FullForm[Hold[a/b]]** Out[3]//FullForm= Hold[Times[a, Power[b, -1]]]

It is important to keep this in mind when writing patterns for subtractions and divisions. In section 5.7, we have seen that the pattern f[x_ - y_] on the left side of a definition is turned into f[x_ + -1 y_] before the rule is defined. It will therefore match the expression `f[a - b]` as intended. It will, however, not match `f[a - 2b]` or `f[a - 3]`, for example. The internal form of these two expressions are `f[a + -2 b]` and `f[a + -3]`. A negative number -2 is not stored as `-1*2` but rather as a single object. The pattern `f[x_ + -1 y_]` does not match.

Since any evaluated product contains at most one negative number we can use the pattern `f[x_ + n_?Negative y_.]` to match any difference. Note that we make `y_.` optional for the case `f[a + -3]`.

Here are some expressions to test our patterns. All should match a difference, except the last one.

```
In[1]:= e = {a - b, a - 2b, a - 3, -a + b, a -b/3, a + b}
                                                b
Out[1]= {a - b, a - 2 b, -3 + a, -a + b, a - -, a + b}
                                                3
```

The naive pattern matches only symbolic subtractions.

```
In[2]:= Cases[ e, x_ - y_ ]

Out[2]= {a - b, -a + b}
```

Allowing for any negative number matches all cases except for a negative number alone.

```
In[3]:= Cases[ e, x_ + n_?Negative y_ ]
                                      b
Out[3]= {a - b, a - 2 b, -a + b, a - -}
                                      3
```

This matches all cases we want.

```
In[4]:= Cases[ e, x_ + n_?Negative y_. ]
                                          b
Out[4]= {a - b, a - 2 b, -3 + a, -a + b, a - -}
                                          3
```

For division we have to deal with a negative exponent instead of a negative factor. The pattern `f[x_/y_]` is turned into `f[x_ y_^-1]` before the rule is defined. It will therefore match `f[a/b]` as intended. It will, however, not match `f[a/b^2]`, for example. The internal form of this expression is `f[a b^-2]` and the pattern `f[x_ y_^-1]` does not match. Instead we should use the pattern `f[x_. y_^n_?Negative]` to match any quotient. Note that we make `x_.` optional for the case `f[1/b]` which is stored as `f[b^-1]`.

Here are some expressions to test our patterns. All should match a quotient, except the last one.

```
In[1]:= e = {a/b, a/b^2, a/Sqrt[b], 1/b, a b}
             a   a      a       1
Out[1]= {-,  --,  -------,  -,  a b}
             b   2   Sqrt[b]   b
                 b
```

The naive pattern matches only symbolic divisions.

```
In[2]:= Cases[ e, x_/y_ ]
            a
Out[2]= {-}
            b
```

Allowing for any negative exponent matches all cases except for a reciprocal.

```
In[3]:= Cases[ e, x_ y_^n_?Negative ]
```

$$Out[3]= \{\frac{a}{b}, \frac{a}{b^2}, \frac{a}{Sqrt[b]}\}$$

This matches all cases we want.

```
In[4]:= Cases[ e, x_. y_^n_?Negative ]
```

$$Out[4]= \{\frac{a}{b}, \frac{a}{b^2}, \frac{a}{Sqrt[b]}, \frac{1}{b}\}$$

The pattern `x_. y_^n_?Negative` does not match a rational number. Although a rational number prints as a fraction p/q, it is a single object in *Mathematica*. To match a rational number use `r_Rational` and refer to its denominator with `Denominator[r]`.

`x_ + n_?Negative y_.`	matching any difference
`{x, -n y}`	referring to the two terms
`x_ y_.^n_?Negative`	matching any quotient
`{x, y^-n}`	referring to the two terms
`r_Rational`	matching a numerical fraction
`{Numerator[r], Denominator[r]}`	referring to the two terms

Patterns for subtractions and divisions

■ 6.5 Traversing Expressions

In this section we look at functions that interact with the syntax of *Mathematica*'s expressions. We are only concerned with the *internal* form of expressions, the one into which all input is first translated (by the so-called *parser*).

■ 6.5.1 The Syntax of Expressions

A computer language needs a precise *syntax*, a set of rules that describe all formally correct expressions. You as a user of a programming language need to know how to write down your input and the parser for the language must be able to uniquely recognize your input.

A language is usually defined by two concepts. The first is that of the building blocks of all expressions, the things that cannot be taken apart further, sometimes called *atoms*. In *Mathematica* these atoms are the symbols, numbers, and strings.

The second concept gives the rules for making more complicated expressions from simpler ones. In *Mathematica* there is only one such rule. Given expressions e_0, e_1, ..., e_n for $n \geq 0$ then the following is also an expression:

$$e_0[e_1, \ldots, e_n].$$

If $n = 0$ this looks like $e_0[\ ]$. All expressions are built up in this way by starting with some atoms and using the above rule many times. e_0 is called the *head* of the expression and the e_1, ..., e_n are called the *elements*.

For example, let us try to understand how the expression

```
Derivative[1][f][x]
```

is constructed. In this case n is 1 and e_0 is the expression `Derivative[1][f]` and e_1 is the expression `x`. `x` is a symbol and we have reached an atom. `Derivative[1][f]` is again built up according to our rule, with $n = 1$, e_0 being `Derivative[1]` and e_1 being the symbol `f`. We need to apply the rule one more time to `Derivative[1]`. e_0 is the symbol `Derivative` and e_1 is the number 1. We have now completely decomposed the expression into its building blocks. This decomposition is unique. There is no other way we could have applied the rule.

■ 6.5.2 Defining Your Own Language

You can define your own rules for a subset of *Mathematica*'s expressions. The set of all expressions that satisfy the rules is technically called a *language*. The rules are called a *grammar*.

For an example, we define a grammar for "algebraic expressions." We have to define the atoms and the rules for combining algebraic expressions to new ones.

The atoms of algebraic expressions are:

- Integer numbers are algebraic expressions.
- Rational numbers are algebraic expressions.
- Complex numbers with integer or rational parts are algebraic expressions.
- Symbols are algebraic expressions.

Given algebraic expressions $e_1, \ldots, e_n$, the following are also algebraic expressions:

- Plus[$e_1, \ldots, e_n$] (the sum of algebraic expressions).
- Times[$e_1, \ldots, e_n$] (the product of algebraic expressions).
- Power[e_1, r], where r is an integer or rational number (rational powers of algebraic expressions).

By writing expressions in their internal forms, you can convince yourself that the examples in the following table are all algebraic expressions.

1 - x	Plus[1, Times[-1, x]]
Sqrt[a + 1]	Power[Plus[a, 1], 1/2]
x/y	Times[x, Power[y, -1]]
I	Complex[0, 1]

Examples of algebraic expressions

■ 6.5.3 Recognizing a Language

Having defined a grammar for algebraic expressions, we now want to be able to *recognize* them, i.e., to find out whether a given expression is an algebraic expression. For this purpose, we define a predicate AlgExpQ[*expr*] that returns True or False depending on whether *expr* is an algebraic expression. This is very easy. We can give definitions for AlgExpQ[] that correspond to all of the rules in our grammar.

First the rules that define the atoms. We use pattern matching for the types of atoms that are algebraic expressions. We give one rule for each kind of atom.

```
AlgExpQ[ _Integer ]  = True
AlgExpQ[ _Rational ] = True
AlgExpQ[ c_Complex ] = AlgExpQ[Re[c]] && AlgExpQ[Im[c]]
AlgExpQ[ _Symbol ]   = True
```

Rules for atoms

The rule for complex numbers works because the parts of a complex number are always other numbers and so the only cases that could match are the first two rules. There is a certain ambiguity in complex numbers. We could have defined them as composite expressions of the form `Complex[re, im]`. It is better to treat them as atoms, however.

The rules for composite algebraic expressions are by nature recursive. In order for a sum of algebraic expressions to be an algebraic expression, *all* of its terms must be algebraic expressions. Therefore we use the logical *and* `&&` on the right side. It is sufficient to give rules for sums and products of *two* terms as we have seen in subsection 6.4.2.

```
AlgExpQ[ a_ + b_ ] := AlgExpQ[a] && AlgExpQ[b]
AlgExpQ[ a_ * b_ ] := AlgExpQ[a] && AlgExpQ[b]
AlgExpQ[ a_ ^ b_Integer ]  := AlgExpQ[a]
AlgExpQ[ a_ ^ b_Rational ] := AlgExpQ[a]
```

Rules for composite algebraic expressions

Finally, we need a rule that matches if none of the above does and returns `False` since everything else is *not* an algebraic expression.

```
AlgExpQ[_] = False
```

Catchall for algebraic expressions

The predicate should return `True` for all the examples given earlier.

```
In[2]:= AlgExpQ[1 - x]
Out[2]= True
```

We have made `AlgExpQ[]` listable to test a whole list of expressions.

```
In[3]:= AlgExpQ[{Sqrt[a + 1], I^I, Sqrt[-1]}]
Out[3]= {True, False, True}
```

Mathematica evaluates the argument of `AlgExpQ[]` in the normal way. Even though what we typed in is not an algebraic expression according to our grammar, it evaluates to -1 which certainly is.

```
In[4]:= AlgExpQ[ Exp[I Pi] ]
Out[4]= True
```

■ 6.5.4 Splitting Atoms

We know now that atoms are not the ultimate building blocks of the universe they were believed to be when the term *atom* was used to denote the smallest parts from which a language is built up. In most programming languages, the atoms can be manipulated in some way too. The functions that do this operate outside of the syntax since in the syntax given, the atoms *are* the fundamental units. To manipulate an atom (a symbol or a number), we can convert it to a string and then we convert the string into a list of its characters. Now we can use the normal *Mathematica* operation to manipulate this list and then we convert things back to an atom.

These concepts are taken from the programming language LISP, with which *Mathematica* shares many common concepts. The function Explode[*symbol*] turns a symbol into a list of characters that make up its name and Intern[*charlist*] is its inverse, converting a list of characters back into a symbol.

```
Explode::usage = "Explode[expr] turns an expression into
    a list of characters that make up its name."

Intern::usage = "Intern[charlist] turns a list of characters into an expression."

Begin["`Private`"]

Explode[atom_] := Characters[ ToString[InputForm[atom]] ]

Intern[l:{_String..}] := ToExpression[ StringJoin @@ l ]

End[]

Protect[Explode, Intern]

Null
```

Atoms.m:Converting expressions to lists of characters

Incidentally these functions work for any *Mathematica* expression, not just atoms. The pattern in the definition of Intern[] matches any list whose elements are all strings. There is a short section on repeated patterns of the form *pattern*.. in section 2.3.8 of the *Mathematica* book.

We get a list of the characters of the name Explode.	`In[2]:= Explode[ Explode ]` `Out[2]= {E, x, p, l, o, d, e}`
We need to look at the input form of it to see that the elements of the list are indeed strings.	`In[3]:= InputForm[ % ]` `Out[3]//InputForm= {"E", "x", "p", "l", "o", "d", "e"}`
The functions Explode and Intern are inverses. The result of applying one to the result of the other is the original expression.	`In[4]:= Intern[ Explode[ symbol ] ]` `Out[4]= symbol`

You can amuse yourself nesting Explode[].	`In[5]:= Explode[ Explode[ ab ] ] // InputForm`
	`Out[5]//InputForm=`
	`   {"{", "\"", "a", "\"", ",", " ", "\"", "b", "\"", "}"}`

Chapter 7

Numerical Computations

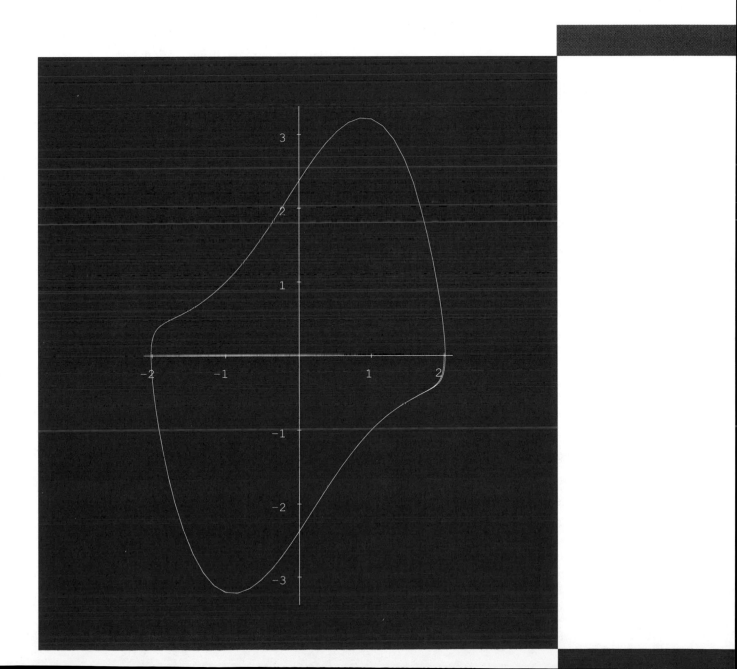

Mathematica has a way of dealing with numerical computations that is not available in most other programming languages: numbers can be of any size and precision. Whenever possible, *Mathematica* maintains expressions in exact form. It does not go to a decimal approximation of `Sqrt[2]`, for example, unless you tell it to.

Section 1 is about the different kinds of numbers *Mathematica* can deal with. It explains the concepts of precision and accuracy that are important for approximate numbers.

In section 2, we look at numerical evaluation (the important command `N[]`). We show how you can define your own numerical procedures.

Section 3 is an application of the material so far and looks at numerical integration of differential equations. The package developed here can deal with simple cases and shows the principles involved. Writing a general-purpose numerical integrator is beyond the scope of this book. It requires advanced knowledge of numerical mathematics. If you do have that knowledge, you should find it easy to write numerical programs in *Mathematica*.

About the illustration overleaf:

A phase-space plot of the *Van-der-Pol equation* $\ddot{x}+x = \epsilon(1-x^2)\dot{x}$ with $\epsilon = 1.5$, integrated numerically.

```
<< RungeKutta.m
eps = 1.5
RungeKutta[{xdot, eps(1-x^2)xdot - x}, {x, xdot}, {2, 0}, {4Pi, 0.05}]
ListPlot[%, PlotJoined -> True, AspectRatio -> Automatic]
```

■ 7.1 Numbers in *Mathematica*

Mathematica supports four different kinds of numbers: integers, rational numbers, complex numbers, and approximate numbers. The real and imaginary part of a complex number can be any of the other three number types. The numerator and denominator of rational numbers are integers. The first two types of numbers and the integer or rational complex numbers are *exact* numbers. You do not lose accuracy when doing computations with them. Computations with approximate numbers or complex numbers with approximate components can only be performed to a certain accuracy and precision and a long sequence of calculations can lose accuracy.

There is no built-in limit on the size or number of digits of numbers in *Mathematica*. Ultimately, the limits will be your computer's memory size and your patience.

■ 7.1.1 Rational and Complex Numbers

Rational numbers and complex numbers each consist of two parts. Nevertheless they are treated as single objects for most operations, including pattern matching.

A complex number is printed as *a + b* I. *a* is called the *real part*, *b* the *imaginary part*. Internally this complex number is stored as Complex[*a, b*]. The symbol I does not occur in this representation! This is important for pattern matching. The pattern x_ + I y_ cannot be used for matching complex numbers. Instead, use z_Complex and use Re[z] and Im[z] on the right side of the rule for x and y.

This is the internal form of complex numbers.	`In[1]:= FullForm[ 2 + 3 I ]` `Out[1]//FullForm= Complex[2, 3]`
Rules like this do not match.	`In[2]:= % /. x_ + I y_ :> {x, y}` `Out[2]= 2 + 3 I`
This is how patterns for complex numbers should be used.	`In[3]:= %% /. z_Complex :> {Re[z], Im[z]}` `Out[3]= {2, 3}`
The symbol I itself is turned into a complex number.	`In[4]:= FullForm[ I ]` `Out[4]//FullForm= Complex[0, 1]`
Again this method of replacing imaginary parts by referring to the symbol I does not work in general.	`In[5]:= f[I] + g[2I] + h[1 - I] /. I -> 0` `Out[5]= f[0] + g[2 I] + h[1 - I]`
Setting I equal to 0 is equivalent to taking the real part of a complex number.	`In[6]:= f[I] + g[2I] + h[1 - I] /. z_Complex :> Re[z]` `Out[6]= f[0] + g[0] + h[1]`

Similar considerations apply to rational numbers. They are printed as *num/den* but are represented as Rational[*num*, *den*]. For pattern matching you cannot use the pattern a_/b_. Use q_Rational and refer to a and b as Numerator[q] and Denominator[q].

■ 7.1.2 Approximate Numbers

Floating-point numbers are described by two quantities, their *accuracy* and their *precision*. The precision is the total number of significant digits and the accuracy describes the position of the decimal point within these digits. A number with precision p has p significant digits, in the figures below denoted by d_1, d_2, ..., d_p. *Significant digits* means that d_1 is nonzero. The accuracy is denoted by a. Depending on the value of a, there are three different cases

$$(1) \qquad 0 < a < p, r = p - a \qquad \overbrace{d_1 \ldots d_r . \underbrace{d_{r+1} \cdots d_p}_{a}}^{p}$$

$$(2) \qquad a \geq p \qquad \underbrace{0 . 0 \cdots 0 \overbrace{d_1 \cdots d_p}^{p}}_{a}$$

$$(3) \qquad a \leq 0 \qquad \overbrace{d_1 \cdots d_p}^{p} \underbrace{0 \cdots 0}_{-a} . 0 .$$

In case 1, the accuracy is less than the precision and there are digits to both sides of the decimal point. In case 2, all digits are to the right of the decimal point and there are $r = a - p$ leading zeroes (we have used this formula in subsection 4.3.4). In case 3, all digits are to the left of the decimal point. The accuracy is *negative*! $r = -a$ digits are undetermined and are therefore set to zero. Exact numbers have a precision and accuracy of Infinity.

In all cases $p - a$ gives the number of digits to the left of the decimal point (this will be negative in case 2) and a itself is, of course, the number of digits to the right of the decimal point (negative in case 3).

This auxiliary function reports both precision and accuracy of numbers.

```
In[1]:= pa[r_?NumberQ] := {Precision[r], Accuracy[r]}
```

This is case 1.

```
In[2]:= pa[ 5234567890.123456789 ]
Out[2]= {19, 9}
```

This is case 2.

```
In[3]:= pa[ 0.005234567890123456789 ]
Out[3]= {19, 21}
```

This is case 3.

```
In[4]:= pa[ 5.234567890123456789 10^30 ]
Out[4]= {19, -12}
```

Internally, precision and accuracy are maintained in bits and not in decimal digits. The results of `Precision[]` and `Accuracy[]` are rounded to the nearest number of decimal digits. Therefore the calculations can be off by one digit.

This number has 19 decimal digits, but because of rounding the result of `Precision[]` is 18.

```
In[5]:= pa[ 0.1234567890123456789 ]
Out[5]= {18, 19}
```

This number also has 19 decimal digits and the rounded result of `Precision[]` agrees.

```
In[6]:= pa[ 0.9234567890123456789 ]
Out[6]= {19, 19}
```

When you enter a number like `1.0`, you will notice that its precision is not 1 or 2 but some higher number (typically 16 or 18). *Low-precision* numbers have a minimum precision, the so-called *machine* precision. For performance reasons, *Mathematica* uses the hardware of the computer it is running on to perform low precision approximate computations. For these numbers it is impossible to keep track of their precision and it is assumed that it is always the same. To find the machine precision of your computer, use the expression `Precision[1.0]`. You should not put its particular value into your program as this may cause it to behave differently on other computers.

High-precision computations are performed in software and it is possible to keep track of every single bit of precision. If you need full control over precision and accuracy, you should perform your calculations with a precision higher than the machine precision.

The number zero is an interesting special case. Its precision is 0 since there are no digits at all. The accuracy of 0.0 on input is taken to be the negative exponent of the smallest positive machine number which therefore depends on the computer used. If 0.0 is the result of some computation, then its accuracy depends on the accuracy of the numbers in this computation.

This is the machine precision of the computer on which this book was formatted.

```
In[7]:= Precision[1.0]
Out[7]= 16
```

Here are precision and accuracy of the approximate number zero on input.

```
In[8]:= pa[0.0]
Out[8]= {0, 323}
```

This 0.0 is the result of a subtraction. The accuracy cannot be more than that of the operands.

```
In[9]:= pa[ N[Sqrt[2], 100]^2 - 2 ]
Out[9]= {0, 99}
```

Exact zero is quite a different matter.

```
In[10]:= pa[0]
Out[10]= {Infinity, Infinity}
```

■ 7.1.3 Combining Numbers

When two approximate numbers are added or multiplied together the precision and accuracy of the result are determined from the precision and accuracy of the operands. They are chosen so as to guarantee that all digits in the result are correctly determined from the digits of the operands.

For multiplication, the *precision* of the result is the minimum of the precisions of the operands. The accuracy is then determined by the position of the decimal point much in the same way as with long multiplication.

The product of two numbers with precision 100 and 40, respectively, has precision 40.

```
In[1]:= pa[ N[Pi, 100] N[E, 40] ]
Out[1]= {40, 39}
```

Multiplying by a machine number gives a machine number.

```
In[2]:= pa[1.0 N[Pi, 100]]
Out[2]= {16, 15}
```

For addition, the *accuracy* of the result is the minimum of the accuracies of the operands. The precision is then determined by the number of digits in the result. The technique is the same as in long addition.

The sum of two numbers with accuracy 20 and 9, respectively, has accuracy 9.

```
In[3]:= pa[0.01234567890123456789 + 1234567890.123456789]
Out[3]= {18, 9}
```

Since the accuracy of 1.0 is only 16 (machine precision), the 1 in the 20th decimal position is discarded.

```
In[4]:= 1.0 + 10.^-20
Out[4]= 1.
```

If an exact number is combined with an approximate number the same rules apply with the precision and accuracy of the exact number being infinite as we have seen.

The precision of the approximate number is preserved.

```
In[5]:= pa[ 100 N[Sqrt[2], 100] ]
Out[5]= {100, 98}
```

The accuracy of the approximate number is preserved. The precision has increased considerably.

```
In[6]:= pa[ 10^100 + 123456789.0123456789 ]
Out[6]= {110, 10}
```

The last example shows that the precision can increase if we add an approximate number and an exact number. We have seen this in an example in subsection 4.1.4 on page 74, where we performed the calculation x = 1 + 1/x repeatedly. If 1/x is less than 1, the precision of the sum is increased.

There is one important exception to this rule. If a machine number is added to an exact number, the precision is never increased beyond machine precision. This is done for performance reasons.

The accuracy is not increased and we again lose the digit in the 20th decimal position.

```
In[7]:= 1 + 10.^-20
Out[7]= 1.
```

If the inexact number has a higher precision, no accuracy is lost.

```
In[8]:= 1 + N[10^-20, 20]
Out[8]= 1.0000000000000000001
```

The two components of a complex number are always either both exact or approximate. The precision of the complex number is taken to be the maximum precision of its two parts. The accuracy of both parts is set to the minimum of the accuracies.

The accuracy of the real part is too small for the imaginary part to be retained.

```
In[1]:= N[10^10, 20] + N[10^-10, 20] I
Out[1]= 10000000000.
```

This is just enough accuracy.

```
In[2]:= N[10^10, 21] + N[10^-10, 20] I
                             -10
Out[2]= 10000000000. + 1. 10   I
```

■ 7.1.4 Setting Precision and Accuracy

It is possible to "cheat," i.e., to artificially raise precision and accuracy of numbers. This is useful to perform intermediate computations at a higher accuracy. In most cases, the accuracy of the result should then be lowered back to at most the accuracy of the input. We have used this technique in subsection 4.3.4.

This is a machine approximation of 1/3.

```
In[1]:= third = 1./3
Out[1]= 0.333333
```

This increases the precision to 30 digits. The added digits are normally not 0 in base 10 representation.

```
In[2]:= SetPrecision[third, 30]
Out[2]= 0.333333333333333314829616256247
```

The added digits are 0 in base two, the internal representation for numbers.

```
In[3]:= InputForm[BaseForm[%, 2]]
Out[3]//InputForm=
    2^^0.010101010101010101010101010101010101010101010101\
      010100000000000000000000000000000000000000000000000000
```

Of course, you can also *lower* the precision.

```
In[4]:= SetPrecision[1/7, 40]

Out[4]= 0.1428571428571428571428571428571428571429
```

This sets the *accuracy* to 30. Again the extra digits are not 0 in base 10.

```
In[5]:= SetAccuracy[1./9, 30]

Out[5]= 0.111111111111111104943205418749
```

■ 7.1.5 Application: Arithmetic with Mathematical Constants

Mathematica knows about the mathematical constants Pi, Degree, GoldenRatio, E, EulerGamma, and Catalan. None of these can be represented as an exact number and they are consequently left alone when encountered in an expression. If you do arithmetic with such a constant and an approximate number, then it makes sense to convert it to an approximate number and perform the arithmetic operation. *Mathematica* does not do this by itself. It is easy to give some rules for this. Here they are:

```
BeginPackage["Constants`"]

Constants::usage = "Constants.m defines rules for approximate arithmetic with the
    mathematical constants Pi, Degree, GoldenRatio, E, EulerGamma and Catalan."

Begin["`Private`"]

constantList = {Pi, Degree, GoldenRatio, E, EulerGamma, Catalan}

protected = Unprotect[Release[constantList]]

DefineRules[const_Symbol] := (
    const/: const + r_ := N[const, Accuracy[r]] + r /; Accuracy[r] < Infinity;
    const/: const * r_ := N[const, Precision[r]] * r /; Precision[r] < Infinity;
    const/: const ^ r_ := N[const, Precision[r]] ^ r /; Precision[r] < Infinity
    )

DefineRules /@ constantList

Protect[Release[protected]]

End[]

EndPackage[]
```

Constants.m: Performing arithmetic with mathematical constants

These rules use N[] to compute an approximation of the constant to the accuracy or precision needed. They do this whenever the constant occurs in an arithmetic expression with a number whose precision is less than infinity. Note that we use the accuracy for addition and the precision for the other operations. The auxiliary function DefineRules[] defines the rules. In this way, we save ourselves the repetitive labor of typing in all the similar rules for all the constants. These techniques were explained in section 5.5.

`Pi` is not evaluated numerically.	`In[1]:= 1.0 + Pi` `Out[1]= 1. + Pi`
A typical way to set up the needed definitions, especially inside a package.	`In[2]:= Needs["Constants`"]`
Approximate numbers are now "contagious."	`In[3]:= 1.0 + Pi` `Out[3]= 4.14159`
To get the value of a constant itself, simply multiply it by 1.0.	`In[4]:= 1.0 EulerGamma` `Out[4]= 0.577216`

■ 7.2 Numerical Evaluation

There are two circumstances in which *Mathematica* performs approximate numerical computations. If an expression contains approximate numbers then arithmetic involving these numbers and possibly other exact numbers is performed according to the rules we have just seen in section 7.1. If a mathematical function receives an approximate number as argument, then its value is computed by some built-in algorithm and it returns an approximate result.

The second way to perform numerical computation is to apply the command N[] to an expression. This is referred to as *numerical evaluation*.

■ 7.2.1 The Command N

If you apply N[] to an expression in the form N[*expr, prec*], where *prec* is the optional precision with default the machine precision, several things happen.

- N[] applies itself recursively to all subexpressions.

- N[*r, prec*] converts the number *r* to an approximate number with precision at most *prec*. It does not increase the precision of a lower precision approximate number.

- N[*const, prec*], where *const* is one of the mathematical constants listed in section 3.2.8 of the *Mathematica* book, computes a numerical approximation to precision *prec*. (See also subsection 7.1.5.)

- For certain functions, N[*func*[*args*..], *prec*] uses a different built-in function than *func*[*args*..] would. These numerical versions of the functions have names of the form N*func*[], for example NIntegrate[]. Note that the argument of N[] is evaluated first, i.e., it first tries to evaluate *func*[*args*..] symbolically. A list of these functions is in section 3.9.1 of the *Mathematica* book.

- For all other expressions, N[*expr, prec*] gives just *expr*.

N[] applies itself to all subexpressions. Here it turns 5 into 5.0 and then the normal code for Sin[] computes the result.

```
In[1]:= N[Sin[5]]
Out[1]= -0.958924
```

Exact numbers are converted to approximate numbers.

```
In[2]:= N[1/3, 30]
Out[2]= 0.333333333333333333333333333333
```

The precision is never increased. You can use SetPrecision[] to do this.

```
In[3]:= N[%, 100]
Out[3]= 0.333333333333333333333333333333
```

Mathematical constants are computed to the required precision.

```
In[4]:= N[Catalan, 40]
Out[4]= 0.9159655941772190150546035149323841110774
```

The function `NSum[]` is used to find the value of this infinite sum.

```
In[5]:= N[Sum[1/i^2, {i, 1, Infinity}]]
Out[5]= 1.64493
```

`Sum[]` alone does not do this, even if its argument is an approximate number.

```
In[6]:= Sum[1.0/i^2, {i, 1, Infinity}]
                 1.
Out[6]= Sum[--, {i, 1, Infinity}]
                  2
                 i
```

`N[]` does nothing in other cases.

```
In[7]:= N[somethingelse]
Out[7]= somethingelse
```

■ 7.2.2 Numerical Procedures

As mentioned in the previous subsection, certain built-in functions have a separate numerical version. There is an important difference between, for example,

$$N[\text{Sum}[expr, iterator]]$$

and

$$\text{NSum}[expr, iterator].$$

The former first evaluates `Sum[`*expr*, *iterator*`]` since the argument of any function, including `N[]`, is evaluated first. Only if the result of this evaluation is still of the form `Sum[`*expr*, *iterator*`]`, does it then call `NSum[]`.

Some of these numerical procedures have a non-standard way of handling precision. Instead of taking the precision from their argument, they use options. Therefore, to increase the precision you cannot use

$$N[\text{Sum}[expr, iterator], prec]$$

but have to use

$$\text{NSum}[expr, iterator, \text{AccuracyGoal} \rightarrow prec, \text{WorkingPrecision} \rightarrow prec{+}extra].$$

These options are explained in the Reference Guide.

The precision is not increased.

```
In[1]:= N[Sum[1/i^2, {i, 1, Infinity}], 20]
Out[1]= 1.64493
```

Now it works. The exact value of this sum is $\pi^2/6$.

```
In[2]:= NSum[1/i^2, {i, 1, Infinity},
            AccuracyGoal -> 20, WorkingPrecision -> 30]
Out[2]= 1.6449340668482264365
```

There are many options that you will have to play with in difficult cases.

```
In[3]:= Options[ NSum ]
Out[3]= {WorkingPrecision -> 16, Terms -> 15,
    ExtraTerms -> 11, WynnDegree -> 1, Method -> Automatic,
    AccuracyGoal -> 6}
```

■ 7.2.3 Defining Your Own Numerical Procedures

To make your own numerical routine `Nf[]` behave like the built-in ones, you define a rule for `N[f[x_]]`. Such a rule is automatically stored with `f`. Since `N[]` takes an optional second argument, your rule should do so as well. The template for a numerical rule therefore looks like this.

```
N[f[x_], prec_:(Precision[1.0])] := Nf[x, prec]

Nf[x_, prec_:(Precision[1.0])] := {x, prec} (* numerical code goes here *)
```

Numerical.m: Defining a numerical rule

The first rule causes expressions of the form `N[f[arg]]` to call the numerical rule `Nf[]`, passing it the argument and the desired precision. The second rule for `Nf[]` is where the numerical computation will be performed. It can be called directly and therefore we make the precision optional. In our template, we merely return the arguments in a list. As we have seen in subsection 7.1.2, `Precision[1.0]` is the machine-independent way of specifying machine-precision, the default for `N[]`.

The first rule is applied to turn the expression into `Nf[1/3, 20]`. `N[]` then makes all numbers approximate.

```
In[2]:= N[f[1/3], 20]
Out[2]= {0.33333333333333333333, 20.}
```

We can call `Nf[]` directly. Note that the numbers are not converted to approximate numbers.

```
In[3]:= Nf[5]
Out[3]= {5, 16}
```

Since numbers are not converted to approximate numbers if we call `Nf[]` directly, the first statement in the body of `Nf[]` should probably be `nx = N[x, prec]`. Thus, a typical outline for the numerical procedure `Nf[]` looks like this.

```
Nf[x_, prec_:(Precision[1.0])] :=
    Block[{nx = N[x, prec]},
        ⋮              (* numerical code goes here *)
    ]
```

A typical numerical procedure

As we have seen in the previous subsection, `N[Sum[expr, iterator], prec]` does not do the expected thing, since `NSum[]` wants the precision passed as an option. In numerically uncritical cases, the following numerical rule can be given to automatically pass the precision in the required way.

```
Unprotect[Sum]
N[Literal[Sum[args__]], prec_] :=
    NSum[args, AccuracyGoal->prec, WorkingPrecision->prec+10]
```

Passing the precision to a numerical routine

With the above definition in effect, NSum[] computes its result to the requested precision.

```
In[2]:= N[Sum[1/i^3, {i, Infinity}], 40]

Out[2]= 1.2020569031595942853997381615114499990765
```

The value of this infinite sum is equal to $\zeta(3)$.

```
In[3]:= N[Zeta[3], 40]

Out[3]= 1.2020569031595942853997381615114499990765
```

■ 7.3 Application: Solution of Differential Equations

Mathematica is no replacement for specialized numerical codes that are highly optimized to perform machine-precision calculations. But the merging of numerical and symbolic computation possible in *Mathematica* makes it very well suited for expressing numerical algorithms, testing them out and postprocessing the results, for example in graphical form. Its ability for high-precision or exact calculations is also important.

In this section, we want to look at a program that combines functional programming with numerical computations for numerically integrating systems of ordinary, first-order differential equations with the *Runge-Kutta* method.

The first subsection introduces the mathematical background of the Runge-Kutta method. Its understanding is not necessary for the rest of this section, however.

■ 7.3.1 The Runge-Kutta Method

An autonomous first-order system of differential equations is given by a vector of n functions f_1, f_2, ..., f_n of n variables y_1, y_2, ..., y_n and is of the following form:

$$\begin{aligned}
\dot{y}_1 &= f_1(y_1, y_2, \ldots, y_n) \\
\dot{y}_2 &= f_2(y_1, y_2, \ldots, y_n) \\
&\;\;\vdots \\
\dot{y}_n &= f_n(y_1, y_2, \ldots, y_n)
\end{aligned}$$

where $\dot{y}$ denotes differentiation with respect to the independent variable t.

A solution is a vector $y_1(t)$, $y_2(t)$, ..., $y_n(t)$ of n functions of t that satisfy the given equations and also an initial condition a_1, a_2, ..., a_n at time t_0

$$\begin{aligned}
y_1(t_0) &= a_1 \\
y_2(t_0) &= a_2 \\
&\;\;\vdots \\
y_n(t_0) &= a_n \, .
\end{aligned}$$

An autonomous system is one in which the functions f_1, f_2, ..., f_n do not depend on t. In this case we can take $t_0 = 0$.

Writing $\vec{y}$ for y_1, y_2, ..., y_n and $\vec{f}$ for f_1, f_2, ..., f_n we can write the equations and initial condition simply as

$$\begin{aligned}
\dot{\vec{y}} &= \vec{f}(\vec{y}) \\
\vec{y}(t_0) &= \vec{a} \, .
\end{aligned}$$

A numerical method for solving such a system finds the values of $\vec{y}$ at a number of values of t starting from the initial conditions $\vec{a}$. Given the values $\vec{y}^{(0)} = \vec{y}(t_0)$ it finds the values $\vec{y}^{(1)}$ at time $t_0 + dt$, $\vec{y}^{(2)}$ at time $t_0 + 2dt$ and so on. dt is called the *step size*.

There are many different formulae for this purpose. They differ in the number of evaluations of the functions $\vec{f}$ that are necessary. *Higher-order* methods require many evaluations of $\vec{f}$, but they can find accurate solutions with a larger step size dt and so fewer steps are necessary to find $\vec{y}(t)$ for some given time t.

The fourth order Runge-Kutta formula finds $\vec{y}^{(i+1)}$ given $\vec{y}^{(i)}$ as follows.

$$\vec{k}_1 = dt\,\vec{f}(\vec{y}^{(i)})$$

$$\vec{k}_2 = dt\,\vec{f}(\vec{y}^{(i)} + \frac{\vec{k}_1}{2})$$

$$\vec{k}_3 = dt\,\vec{f}(\vec{y}^{(i)} + \frac{\vec{k}_2}{2})$$

$$\vec{k}_4 = dt\,\vec{f}(\vec{y}^{(i)} + \vec{k}_3)$$

$$\vec{y}^{(i+1)} = \vec{y}^{(i)} + \frac{\vec{k}_1 + 2\vec{k}_2 + 2\vec{k}_3 + \vec{k}_4}{6}.$$

■ 7.3.2 Programming the Runge-Kutta Formula

The formula for the Runge-Kutta method can be programmed in *Mathematica* rather easily. From the beginning we want to make it as flexible as possible. We use lists for the vectors $\vec{f}$ and $\vec{y}$. We can write the code in a way that is completely independent of the number of equations n.

```
RKStep[f_, y_, y0_, dt_] :=
    Block[{ k1, k2, k3, k4 },
        k1 = dt N[ f /. Thread[y -> y0] ];
        k2 = dt N[ f /. Thread[y -> y0 + k1/2] ];
        k3 = dt N[ f /. Thread[y -> y0 + k2/2] ];
        k4 = dt N[ f /. Thread[y -> y0 + k3] ];
        y0 + (k1 + 2 k2 + 2 k3 + k4)/6
    ]
```

The code for one step with the Runge-Kutta formula

The parameters of `RKStep[]` are the list `f` of expressions describing the functions $\vec{f}$, the list `y` of the names of the variables $\vec{y}$, the list `y0` of initial conditions $\vec{y}^{(0)}$, and the step size dt.

To compute the values $\vec{f}(\vec{y}^{(0)})$ we have to substitute the elements of `y0` for the variables `y` in the expressions `f`. In *Mathematica*, this is done with a list of rules like this

$$f \;/.\; \{y_1 \to y0_1,\; y_2 \to y0_2, \ldots,\; y_n \to y0_n\}$$

where y_i is the ith variable and $y0_i$ is the ith initial value. What we are given as parameters is the list of variables and the list of initial conditions. The function `Thread[]` converts the expression y -> y0 or

$$\{y_1,\; y_2, \ldots,\; y_n\} \to \{y0_1,\; y0_2, \ldots,\; y0_n\}$$

into the desired form by interchanging the lists with the rule.

Nowhere do we need to know the length of these lists or the number of variables given. All the arithmetic is performed element by element, since all arithmetic functions are listable. Quite in contrast to a typical numerical code, no loops are needed since *Mathematica*'s language is rich enough to express the underlying ideas directly.

The parameters `f`, `y`, and `dt` are constant for a given system of equations. All we need is a function of a given state y0 that produces the next state at time `dt`. We can turn `RKStep[]` into such a function by using it as function of y0 alone like this: `RKStep[f, y, #, dt]&`. All we have to do now is to iterate this function a suitable number of times. The top level function `RungeKutta[]` does this. Here it is together with the usual package framework. `RungeKutta[]` returns a list of all the $\vec{y}$ values at times 0, dt, $2dt$, ..., $m\,dt$.

```
BeginPackage["RungeKutta`"]

RungeKutta::usage = "RungeKutta[{e1,e2,..}, {y1,y2,..}, {a1,a2,..}, {t1, dt}]
    numerically integrates the ei as functions of the yi with inital values ai.
    The integration proceeds in steps of dt from 0 to t1."

Begin["`Private`"]

RKStep[f_, y_, y0_, dt_] :=
    Block[{ k1, k2, k3, k4 },
        k1 = dt N[ f /. Thread[y -> y0] ];
        k2 = dt N[ f /. Thread[y -> y0 + k1/2] ];
        k3 = dt N[ f /. Thread[y -> y0 + k2/2] ];
        k4 = dt N[ f /. Thread[y -> y0 + k3] ];
        y0 + (k1 + 2 k2 + 2 k3 + k4)/6
    ]

RungeKutta[f_List, y_List, y0_List, {t1_, dt_}] :=
    NestList[ RKStep[f, y, #, N[dt]]&, N[y0], Round[N[t1/dt]] ] /;
        Length[f] == Length[y] == Length[y0]

End[]
EndPackage[]
```

RK1.m: Solving autonomous systems of equations

The number of integration steps is found by dividing `t1` by `dt` and rounding the result.

■ 7.3.3 Example: The Lorenz Attractor

For an example of the use of `RungeKutta[]`, let us look at the equation for the Lorenz Attractor. This is a system of three equations.

$$\dot{x} = -3(x - y)$$
$$\dot{y} = -xz + 26.5x - y$$
$$\dot{z} = xy - z$$

This integrates the Lorenz equations with initial condition $(0, 1, 0)$ from 0 to 20 in steps of size 0.04.

```
In[2]:= RungeKutta[{-3(x-y), -x z + 26.5x - y, x y - z},
                    {x, y, z}, {0, 1, 0}, {20, 0.04} ];
```

The result is a list of points in space which we can plot as a line in three dimensions.

```
In[3]:= Show[ Graphics3D[{Line[%]}], Axes->Automatic ]
```

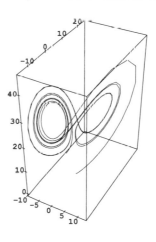

■ 7.3.4 Solving Time-Dependent Systems

In a time-dependent system, the functions $\vec{f}$ depend on t as well as on $\vec{y}$. The equations therefore look like

$$\dot{\vec{y}} = \vec{f}(\vec{y}, t)$$
$$\vec{y}(t_0) = \vec{a}.$$

Rather than writing a completely new procedure for solving such equations, we treat t as an additional dependent variable with the trivial equation $\dot{t} = 1$. We set $y_{n+1} = t$ and $f_{n+1}(y_1, y_2, \ldots, y_{n+1}) = 1$. Having done so, we use the existing code for `RungeKutta[]` to solve it and then remove the last component from all the points in the solution before returning the solution. We do this in a second rule for `RungeKutta[]`.

```
RungeKutta[f_List, y_List, y0_List, {t_, t0_, t1_, dt_}] :=
    Block[{res},
        res = RungeKutta[ Append[f, 1], Append[y, t], Append[y0, t0], {t1 - t0, dt} ];
        Drop[#, -1]& /@ res
    ] /; Length[f] == Length[y] == Length[y0]
```

Part of **RK2.m**: Solving time-dependent systems

For an example, we look at a forced oscillation given by the equation $\ddot{x} + x = -\alpha\dot{x} + \epsilon\cos(\omega t)$. This is one second-order equation. We can transform it into two first-order equations with the variables $x_1 = x, x_2 = \dot{x}$. The equations become

$$\begin{aligned}
\dot{x}_1 &= x_2 \\
\dot{x}_2 &= -x_1 - \alpha x_2 + \epsilon\cos(\omega t).
\end{aligned}$$

In this example, $\alpha = 0.01, \epsilon = 0.1$ and $\omega = 1.1$. The initial condition is $(2, 0)$ and time goes from 0 to 24π in steps of $\pi/18$.

```
In[1]:= RungeKutta[{x2, - x1 - 0.01 x2 + 0.1 Cos[1.1 t]},
                   {x1, x2}, {2, 0},
                   {t, 0, 24Pi, Pi/18} ];
```

The result is a list of points in the plane which we can plot as a line in two dimensions. The solution oscillates around the origin with varying amplitude.

```
In[2]:= Show[ Graphics[{Line[%]}],
              Axes->Automatic, AspectRatio->Automatic ]
```

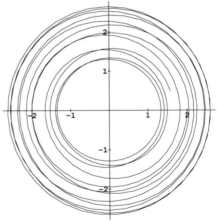

Chapter 8

Overriding Built-in Rules

In this chapter we look at the relationship between built-in rules and methods and user-defined ones. If you want to change the standard behavior of *Mathematica* you can give your own rules that override or augment system definitions. We also look at how you can make your own functions behave like built-in ones.

Section 1 looks a the two different kinds of definitions that can be made for a symbol, the so-called upvalues and downvalues, and then applies these features for defining rules for arithmetic operators and derivatives.

In section 2, we are concerned with the overall behavior of *Mathematica*, its main evaluation loop. Here too, you can change the way things work by assigning functions to global variables provided for this purpose.

Adding more rules to built-in functions is the topic of section 3. If you do not like the way things are set up, simply change them!

Section 4 is an advanced topic that deals with implementing a completely new function in *Mathematica*. The example we chose is one of the few special functions not already built-in. Programming the numerical values for a new function particularly requires advanced knowledge in numerical mathematics. However, the concepts involved in setting up the rules are easier to explain and you can skip the parts of this sections that you are not familiar with.

About the illustration overleaf:

The hyperbolic icosahedron. In hyperbolic geometry the straight lines become circles intersecting an imaginary sphere enclosing the icosahedron at right angles.

■ 8.1 User-Defined Rules Take Precedence

As explained in section B.4 of the Reference Guide, any definitions made by the user are applied before any built-in code is executed to evaluate an expression. It is therefore possible to give rules that override built-in ones. Since the built-in rules will eventually be applied anyway, doing so can be a bit tricky.

■ 8.1.1 Upvalues and Downvalues

A definition of the form

$$f[args...] := body$$

is often called a *downvalue* for f because it defines a value for f appearing at the top level of the rule. A definition of the form

$$f/: g[[f[args...], ...] := body$$

is called an *upvalue* for f. Upvalues are applied before downvalues. We have used upvalues for example in subsection 6.3.1 to give rules for products of trigonometric functions in the form Sin/: Sin[x_] Cos[y_] := body, which *Mathematica* would otherwise leave alone.

■ 8.1.2 Definitions for Arithmetic Operations

Using upvalues for definitions involving arithmetic operations is preferred in the cases where the rules belong to a certain class of expressions, here the trigonometric expressions. If the definition is for a broader class of expressions, not characterized by a certain symbol appearing as their head, this is not possible. For an example, let us develop a rule that will cause all products to expand automatically. Our first try might be

$$a\_ \ b\_ := Expand[a \ b]$$

which will lead to an infinite loop since the argument of the right side is evaluated before Expand[] is called and this argument is again a product.

The solution is to perform the expansion ourselves instead of calling Expand[]. The expansion of the term $(a + b)c$ is $a \ c + b \ c$ and this is all we need to expand any product:

$$(a\_ + b\_)c\_ := a \ c + b \ c.$$

Since we will define a rule for multiplication, we need to unprotect Times[].	In[1]:= Unprotect[Times] Out[1]= {Times}
This definition will expand all products.	In[2]:= (a_ + b_)c_ := a c + b c
It is used repeatedly until no terms with embedded addition remain.	In[3]:= x (u + v + w)(a - 1) Out[3]= -(u x) + a u x - v x + a v x - w x + a w x

With long sums as arguments, this rule will be quite slow. We encountered the same problem with our rules for $\sin(nx)$ in subsection 6.2.3. The rule will be applied many times, each time expanding only one of the terms in the sum. We can use `Map[]` to multiply all the terms in the sum by the second argument of the product in one step:

$$a\_Plus \ c\_ := (\# \ c)\& \ /@ \ a.$$

This rule will be much faster for longer sums.

The expression $a_1 + a_2 + \cdots + a_{100}$ serves as our long sum.	`In[4]:= expr = Array[a, 100, 1, Plus];`
It takes quite a while.	`In[5]:= Timing[expr c;]` `Out[5]= {5.46667 Second, Null}`
This deletes the old rule in preparation for the new one.	`In[6]:= Clear[Times]`
This is the new rule for expansion of products.	`In[7]:= a_Plus c_ := (# c)& /@ a`
It is much faster.	`In[8]:= Timing[expr c;]` `Out[8]= {0.283333 Second, Null}`

Our rule does not expand integer powers of sums. Again we have two choices, a trivial one-step rule and a faster, more complicated formula. The trivial rule reduces exponentiation to the previous case, multiplication:

$$(a\_ + b\_) \wedge n\_Integer?Positive := (a + b)(a + b)\wedge(n-1).$$

In this rule, we see the importance of applying user-defined rules first. The right side of the rule looks like $e \ e \wedge m$ which *Mathematica* would simplify to $e \wedge (m+1)$. Our own rule for multiplying a sum with anything else is applied first, preventing another infinite loop. The faster rule uses the binomial formula $(a+b)^n = \sum_{i=0}^{n} \binom{n}{i} a^i b^{n-i}$ to expand an integer power.

The rule will be defined for `Power` which needs to be unprotected first.	`In[9]:= Unprotect[Power]` `Out[9]= {Power}`
With the rule for expanding products still in effect, we define this rule for expanding powers.	`In[10]:= (a_ + b_) ^ n_Integer?Positive :=` `            (a + b)(a + b)^(n-1)`
Again rather slow.	`In[11]:= Timing[(x + y + z)^10;]` `Out[11]= {5.61667 Second, Null}`

We get rid of this rule.	`In[12]:= Clear[Power]`

Now we replace it with the faster one. The iterator variable has to be isolated in a block since the rule will be used recursively.

```
In[13]:= (a_ + b_) ^ n_Integer?Positive :=
            Block[ {i}, Sum[ Binomial[n, i] a^i b^(n-i),
                      {i, 0, n}] ]
```

This looks much better.

```
In[14]:= Timing[(x + y + z)^10;]

Out[14]= {1.21667 Second, Null}
```

In subsection 8.2.1 we will see a quite different way to achieve the same objective.

■ 8.1.3 Derivatives

The derivative operator `Derivative[n]` represents the nth derivative. It takes a function f as argument and returns another function—the nth derivative $f^{(n)}$. Applying this function $f^{(n)}$ to an argument x gives $f^{(n)}(x)$, the nth derivative at x. In *Mathematica*, this is written as `Derivative[n][f][x]`. If the derivative of f is known, then `Derivative[n][f]` returns a pure function representing that derivative, as we have seen in subsection 5.1.2.

We can define a value for a derivative with `f' = fp` or `Derivative[1][f] = fp`. The head of the left side is `Derivative[1]` which is not a symbol. The rule can therefore not be stored with the head as is normally the case. It is stored with the head of the head, the symbol `Derivative`. It is usually better, however, to store it with the argument `f` with `f/: f' = fp`.

If we do not have a name for the derivate, we can make a definition for `f'[x]`, for example `f'[x_] := 1 + x^2`. The left side is `Derivative[1][f][x_]`. The symbol `f` is not an *argument* of this and we cannot store the rule with `f`, but it is stored with `Derivative`. `Derivative` is one of very few system symbols that are not protected, so rules like these can be given easily.

Since the derivative of tan is known, it is returned as a pure function.	`In[1]:= Tan'`
	
	`Out[1]= Cos[#1]`$^{-2}$` &`

The formal derivative of `f` is set to `g`.	`In[2]:= f' = g`
	
	`Out[2]= g`

The rule is used to simplify the third derivative of `f` at `x`.	`In[3]:= f'''[x]` `Out[3]= g''[x]`

For another example of defining derivatives for known functions see subsection 8.3.

■ 8.2 Modifying the Main Evaluation Loop

Normally, *Mathematica* evaluates what you type in and then prints the value of the evaluated expression. You can modify how *Mathematica* evaluates your input or how it prints the results. This is done by assigning functions to any of the global variables $Pre, $Post, or $PrePrint. You can also redirect input and output to files or even programs instead of the terminal you are using.

■ 8.2.1 Pre- and Post-Evaluation

If $Pre has a value, *Mathematica* evaluates the expression $Pre[*input*] where *input* is the expression that was typed in. For example, with $Pre = Expand *every* expression you type will be expanded.

Expand	expand all expressions
Together	put all expressions over a common denominator
TrigExpand	put trigonometric expressions into normal form
PrintTime	print timing for all evaluations (see subsection 5.3.1)
N	evaluate all expressions numerically
N[#, 100]&	evaluate to 100 digits precision

Some functions to use for $Pre

If $Post has a value, *Mathematica* first evaluates your input in the normal way and then goes through another evaluation with the expression $Post[*val*] in which *val* is the value of the first evaluation.

Share	minimize storage after each evaluation
Expand, Together, ...	same as with $Pre above

Some functions to use for $Post

If the function you use for $Pre or $Post evaluates its argument in the normal way, it does not make a difference whether you assign it to $Pre or $Post since in both cases your input is evaluated first. Assigning a function that does not evaluate its argument to $Post is probably useless.

If $PrePrint has a value, *Mathematica* first evaluates your input in the normal way and assigns the result to the next Out[*n*]. It then evaluates $PrePrint[*val*] and prints the result. A typical choice for $PrePrint is Short which will prevent *Mathematica* from accidentally trying to print pages and pages of output if you are working with long expressions. Another common choice is InputForm which will print all output in a form that is suitable to be pasted back into *Mathematica*.

Short	print only a one line summary
Short[#, 5]&	print five line summary
InputForm	print in input form

Some functions to use for $PrePrint

■ 8.2.2 Application: An Invisible Timing Command

In subsection 5.3.1, we have encountered the function PrintTime[] that prints the time it takes to evaluate its argument and then returns the result of the evaluation. It does therefore not interfere with the normal flow of the computation as Timing[] does. In the previous subsection, we saw that we can assign PrintTime to $Pre to print the time for the evaluation of all future computations in our *Mathematica* session.

For many applications this will be good enough. We can, however, develop a fully transparent version of PrintTime[] called ShowTime[] that has the following additional properties:

- It does not use up $Pre. You can still use $Pre for another purpose.
- Timing information can be turned on and off with the commands On[ShowTime] and Off[ShowTime].

The key idea for reusing $Pre is to remember the value of $Pre before we assign ShowTime to it and to restore that value during the evaluation of the user's input. This is done by binding $Pre as a local variable in a block inside ShowTime. The code contains quite a few subtleties and we will have a closer look at it.

```
ShowTime::usage = "On[ShowTime] turns timing info on. Off[ShowTime] turns it off again.
    The time taken for each command is printed before the result (if any)."

Begin["ShowTime`"]

`oldPre =.          (* user's value of $Pre *)
`ison = False       (* whether ShowTime is currently turned on *)

Attributes[ShowTime] = HoldFirst

ShowTime::twice = "ShowTime is already on."
ShowTime::off = "ShowTime is not in effect."

ShowTime[expr_] :=
```

```
    Block[{`timing},
        Block[{$Pre = oldPre},
            timing = Timing[ oldPre[expr] ];
            Print[ timing[[1]] ];
            oldPre = If[ ValueQ[$Pre], $Pre, Identity ];
        ];
        If[ !ison, $Pre = oldPre ];  (* turn it off again *)
        timing[[2]]
    ]
ShowTime/: On[ShowTime] := (
    If[ ison,
        Message[ShowTime::twice],
    (* else *)
        oldPre = If[ ValueQ[$Pre], $Pre, Identity ];
        $Pre = ShowTime;
        ison = True
    ]; )
ShowTime/: Off[ShowTime] := (
    If[ ison,
        ison = False,
    (* else *)
        Message[ShowTime::off]
    ]; )
End[]

On[ShowTime]
```

ShowTime.m: An improved version of **PrintTime.m**

Let us first look at how things are initially set up when On[ShowTime] is executed. If ShowTime has already been turned on, we can print an error message and do nothing. Otherwise, we assign the current value of $Pre to the local variable oldPre. This is not quite straightforward since $Pre need not have a value at all. In this case we set the value of oldPre to Identity which is as good as nothing at all. Now we assign ShowTime to $Pre and remember that it has been turned on in the local variable ison.

The next time an expression *expr* is evaluated, *Mathematica* will evaluate the expression ShowTime[*expr*] since the value of $Pre is ShowTime. ShowTime does not evaluate its argument and so the evaluation starts inside the body of ShowTime. The outer block declares a local variable that is used to hold the result of the call to Timing[] that is to follow. This is similar to PrintTime[]. Before we call Timing[] on the user's input *expr* we do two things. First we restore the old value of $Pre by declaring $Pre as a local variable to the inner block and initializing it with the saved value oldPre. Any variable, including system symbols, can be localized. Then we apply the old value `oldPre` to *expr* before evaluating it inside `Timing[]`. In this way any value the user has defined for `$Pre` will take effect just as if `ShowTime` were not there at all.

The time it took to evaluate the user's input is now printed. Before we leave the inner block, we have to reset the value of `oldPre` since one of the consequences of that evaluation could have been to change the user's idea of the value of `$Pre` (If the user had typed `$Pre = Expand`, for example).

To turn off `ShowTime` the user would have typed `Off[ShowTime]`. This would have happened inside `ShowTime[]` of course. `Off[ShowTime]` simply sets the local variable `ison` to `False` and we test for that inside `ShowTime[]`. If `ison` is `False`, we restore `$Pre` to the saved value `oldPre`. Finally, we return the result of the evaluation of the user's input.

Before reading in **ShowTime.m**, `$Pre` might already have some value.	`In[1]:= $Pre = Together`
	
	`Out[1]= Together`

This reads in the package and also turns ShowTime on.	`In[2]:= << ShowTime.m`

From the user's point of view, `$Pre` still has the old value.	`In[3]:= $Pre`
	`0.`
	
	`Out[3]= Together`

And it works as expected.	`In[4]:= 1/x + x/(1-x) + (1-x)/(1+x)`
	`0.15 Second`
	
	$$Out[4]= \frac{1 + x - 2 x^2 + 2 x^3}{(1 - x)\, x\, (1 + x)}$$

We can remove the user's value of `$Pre` and ShowTime will not be affected.	`In[5]:= $Pre=.` `0.`
Now we turn off `ShowTime`.	`In[6]:= Off[ShowTime]` `0.`
The only difference is that now `$Pre` has the value `Identity` instead of no value at all.	`In[7]:= $Pre` `Out[7]= Identity`

■ 8.3 Adding to a System Function

■ 8.3.1 Adding Additional Rules

Mathematica does not always know about all the mathematical properties of its built-in functions. Often such properties are only valid for a restricted domain (for example, only for real-valued arguments) and since all variables are assumed to be complex-valued, no rules are applied. If you know that you will only work with real variables, you can add the necessary rules yourself. An example is the rules for the simplification of expressions involving Re[] and Im[] defined in the package **ReIm.m**. In this section we want to look at another case, Abs[] and Sign[].

For non-numerical arguments x, *Mathematica* does not simplify Abs[x]. Since the absolute value $|x|$ is equal to x for positive x and equal to $-x$ for negative x, we can use the predicates Positive[x] and Negative[x] to give conditional rules for this type of simplification.

```
Abs[x_] :=  x /; Positive[x]
Abs[x_] := -x /; Negative[x]
```

Simplifying absolute values

■ 8.3.2 Definitions for Derivatives and Integrals

Now we want to teach *Mathematica* about integrals and derivatives of the absolute value and sign functions. The derivative of $|x|$ is simply sgn x and the derivative of the sign function is 0. At $x = 0$ none of these are differentiable, and so we have to make the rules conditional. A built-in rule simplifies Sign[0] to 0 which is undesirable in this context. We override this by setting Sign[0] to Indeterminate.

```
Abs/: Abs' = Sign

Sign[0] = Indeterminate
Sign'[x_] := 0 /; x != 0
Sign'[0] = Indeterminate
```

Derivatives

The integral of sgn x is of course $|x|$ and the integral of $|x|$ can be written as either $\frac{1}{2}x|x|$ or $\frac{1}{2}x^2$ sgn x. The conversion between the two is done by the formula x sgn $x = |x|$.

```
Sign/: x_ Sign[x_] := Abs[x]

Abs /: Integrate[Abs[x_], x_]   := x Abs[x]/2
Sign/: Integrate[Sign[x_], x_] := Abs[x]
```

Integrals

Integrating sgn z twice gives the result we have defined in our rules.

```
In[1]:= Integrate[ Integrate[Sign[z], z], z ]
```

$$Out[1]= \frac{z \; Abs[z]}{2}$$

The second derivative simplifies back to the original expression.

```
In[2]:= D[ %, {z, 2} ]

Out[2]= Sign[z]
```

The second derivative of $|x|$ is left in a symbolic form since it is not yet know whether x is zero or not.

```
In[3]:= Abs''[x]

Out[3]= Sign'[x]
```

The first derivative of $|x|$ at 0 is indeterminate.

```
In[4]:= Abs'[0]

Out[4]- Indeterminate
```

■ 8.4 Advanced Topic: Introducing a New Mathematical Function

■ 8.4.1 Properties of Built-in Functions

Mathematica has very many special functions built-in. In this section, we look at what it takes to add another function to this collection. For a built-in function, the code usually provides the following:

- Special values that are known exactly in terms of other functions.
- Expansion into power series.
- Numerical evaluation to any accuracy.
- Derivatives and indefinite integrals.

Let us look at the Bessel functions $J_n(x)$ or `BesselJ[n, x]` for example.

For $n = 1/2$ a formula for the exact value is given.

```
In[1]:= BesselJ[1/2, x]

          Sqrt[2] Sin[x]
Out[1]=  ----------------
          Sqrt[Pi] Sqrt[x]
```

This is a series representation of $J_1(x)$ at $x = 0$.

```
In[2]:= Series[BesselJ[1, x], {x, 0, 7}]

              3     5       7
         x   x     x       x          8
Out[2]=  - - -- + --- - ----- + O[x]
         2   16   384   18432
```

20 digits of $J_2(17)$

```
In[3]:= N[BesselJ[2, 17], 20]

Out[3]= 0.1583638412385
```

There is a formula for its derivative.

```
In[4]:= D[BesselJ[1, x], x]

          BesselJ[0, x] - BesselJ[2, x]
Out[4]=  -------------------------------
                        2
```

■ 8.4.2 Definitions for a New Function

We can provide much of the same functionality for other functions that are not built-in. We need to know their mathematical properties, series representations, and special values. Formulae are obtained from handbooks of mathematical functions, for example Gradshteyn & Ryzhik or Abramowitz & Stegun. For an example of a function that is not built-in, let us turn to the *Struve functions* $H_\nu(z)$. These functions are closely related to the Bessel functions.

■ 8.4.3 Special Values

First, the special values. The handbooks contain the following formulae:

$$H_{n+\frac{1}{2}}(z) = Y_{n+\frac{1}{2}}(z) + \frac{1}{\pi} \sum_{m=0}^{n} \frac{\Gamma(m+\frac{1}{2})(\frac{z}{2})^{-2m+n-\frac{1}{2}}}{\Gamma(n+1-m)}$$

$$H_{-(n+\frac{1}{2})}(z) = (-1)^n J_{n+\frac{1}{2}}(z).$$

This is turned into the following two definitions:

```
StruveH[r_Rational?Positive, z_] :=
    BesselY[r, z] +
    Sum[Gamma[m + 1/2] (z/2)^(-2m + r - 1)/Gamma[r + 1/2 - m], {m, 0, r-1/2}]/Pi /;
        Denominator[r] == 2

StruveH[r_Rational?Negative, z_] :=
    (-1)^(-r-1/2) BesselJ[-r, z] /; Denominator[r] == 2
```

Special values of Struve functions

Note that we match the index $n + 1/2$ in the form `r_Rational` with the condition `Denominator[r] == 2`. On the right side, n is expressed as $r - 1/2$ or $-r - 1/2$ in the second formula.

■ 8.4.4 Series Expansion

For the series expansion we find the following defining formula for $H_\nu(z)$:

$$H_\nu(z) = \sum_{m=0}^{\infty} (-1)^m \frac{(\frac{z}{2})^{2m+\nu+1}}{\Gamma(m+\frac{3}{2})\Gamma(\nu+m+\frac{3}{2})}.$$

For a power series of order n, we need to include terms up to $m = (n - \nu - 1)/2$. An expression is most easily turned into a series by adding a term `O[z]^(n+1)` to it. *Mathematica* then converts the whole expression into a series. We can pull the factor $(\frac{z}{2})^{\nu+1}$ out of the summation. Here is the definition:

```
StruveH/: Series[StruveH[nu_?NumberQ, z_], {z_, 0, ord_Integer}] :=
    (z/2)^(nu + 1) Sum[ (-1)^m (z/2)^(2m)/Gamma[m + 3/2]/Gamma[m + nu + 3/2],
        {m, 0, (ord-nu-1)/2} ] + O[z]^(ord+1)
```

Power series definition

If needed, other formulae can be developed for series at points other than 0. It might also be useful to have rules for dealing with series given as *arguments* of `StruveH[]`.

■ 8.4.5 Numerical Evaluation

For numerical evaluation, we can use the same formula. We keep adding more terms until the result no longer changes. For this kind of power series with Γ-functions in the denominator this will be accurate enough.

```
StruveH[nu_?NumberQ, z_?NumberQ] :=
    Block[{s=0, so=-1, m=0, prec = Precision[z]},
        While[so != s,
            so = s;
            s += N[(z/2)^(2m+nu+1)/Gamma[m + 3/2]/Gamma[m + nu + 3/2], prec];
            m++
        ];
        s
    ]
```

Numerical evaluation

It should be noted that in general a good numerical definition should include different methods for different ranges of the values of the argument z. Often, it is also possible to estimate the number of terms needed beforehand and then use Sum[] instead of the While[] loop above. For functions with higher values of the index ν, there are often recurrence relations that express their values in terms of functions with lower index. A good source of numerical methods is the book *Numerical Recipes in C* by Press, Flannery, Teukolsky, and Vetterling.

■ 8.4.6 Derivatives and Integrals

For computing derivatives we find the following formula:

$$H_{\nu-1}(z) - H_{\nu+1}(z) = 2H'_\nu(z) - \frac{(\frac{z}{2})^\nu}{\sqrt{\pi}\,\Gamma(\nu + \frac{3}{2})}.$$

This is easily programmed in *Mathematica*:

```
Derivative[0, n_Integer?Positive][StruveH][nu_, z_] :=
    D[ (StruveH[nu-1, z] - StruveH[nu+1, z] + (z/2)^nu/Sqrt[Pi]/Gamma[nu + 3/2])/2,
       {z, n-1} ]
```

Derivatives of Struve functions

No formulae exist to express indefinite integrals in terms of other known functions.

■ 8.4.7 A Performance Improvement

Summation formulae can often be speeded up by using incremental updates of the quantities involved. In our formula for the numerical values,

$$H_\nu(z) = \sum_{m=0}^{\infty} (-1)^m \frac{(\frac{z}{2})^{2m+\nu+1}}{\Gamma(m+\frac{3}{2})\Gamma(\nu+m+\frac{3}{2})}$$

we notice that for each m the numerator $(\frac{z}{2})^{2m+\nu+1}$ can be computed from the previous one by multiplying by $-(\frac{z}{2})^2$ since the exponent increases by two for every term in the sum. The minus sign takes care of the change of signs of alternating terms expressed by the formula $(-1)^m$. The Γ-functions in the denominator have the nice property that $x\Gamma(x) = \Gamma(x+1)$ and so each successive term can be computed from the previous one without computing a single Γ-function! We keep three variables zf, g1, and g2 that hold the terms in the numerator and the two Γ-functions in the denominator and that are updated for each iteration.

```
StruveH[nu_?NumberQ, z_?NumberQ] :=
    Block[{s=0, so=-1, m=0, prec = Precision[z],
           z2 = -(z/2)^2,k1 = 3/2, k2 = nu + 3/2, g1, g2, zf},
        zf = (z/2)^(nu+1); g1 = Gamma[k1]; g2 = Gamma[k2];
        While[so != s,
              so = s;
              s += N[zf/g1/g2, prec];
              g1 *= k1; g2 *= k2; zf *= z2;
              k1++; k2++; m++
        ];
        s
    ]
```

Faster numerical evaluation

■ 8.4.8 Putting Things Together

After having collected all the formulae we need, we put them into a complete package. We also make the function listable like any built-in one. The complete package **Struve.m** is reproduced in Appendix B.

Known special values are expressed in terms of Γ-functions for which in turn there are special values defined.

```
In[2]:= StruveH[1/2, x]
```
$$Out[2]= \frac{Sqrt[2]}{Sqrt[Pi]\ Sqrt[x]} - \frac{Sqrt[2]\ Cos[x]}{Sqrt[Pi]\ Sqrt[x]}$$

Here is an example of a power series for $H_2(x)$.

```
In[3]:= Series[StruveH[2, x], {x, 0, 8}]
```
$$Out[3]= \frac{2\ x^3}{15\ Pi} - \frac{2\ x^5}{315\ Pi} + \frac{2\ x^7}{14175\ Pi} + O[x]^9$$

The first derivative is expressed again in terms of Struve functions according to our definition.

```
In[4]:= D[ StruveH[1, x], x ]
```
$$Out[4]= \frac{\dfrac{2\ x}{3\ Pi} + StruveH[0, x] - StruveH[2, x]}{2}$$

The numerical definitions given allow us to evaluate the functions anywhere and thus also plot them. A similar plot appears in Abramowitz & Stegun.

```
In[5]:= Plot[{StruveH[0, x], StruveH[1, x],
              StruveH[2, x]},
             {x, 0, 13}, PlotRange->{-0.3, 2}]
```

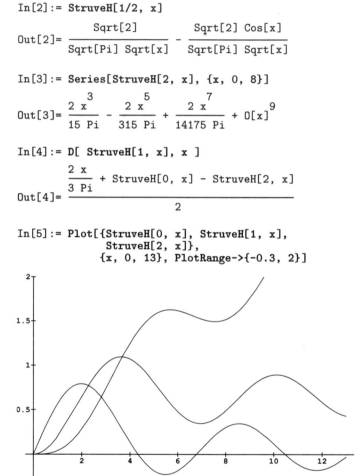

Chapter 9

Input and Output

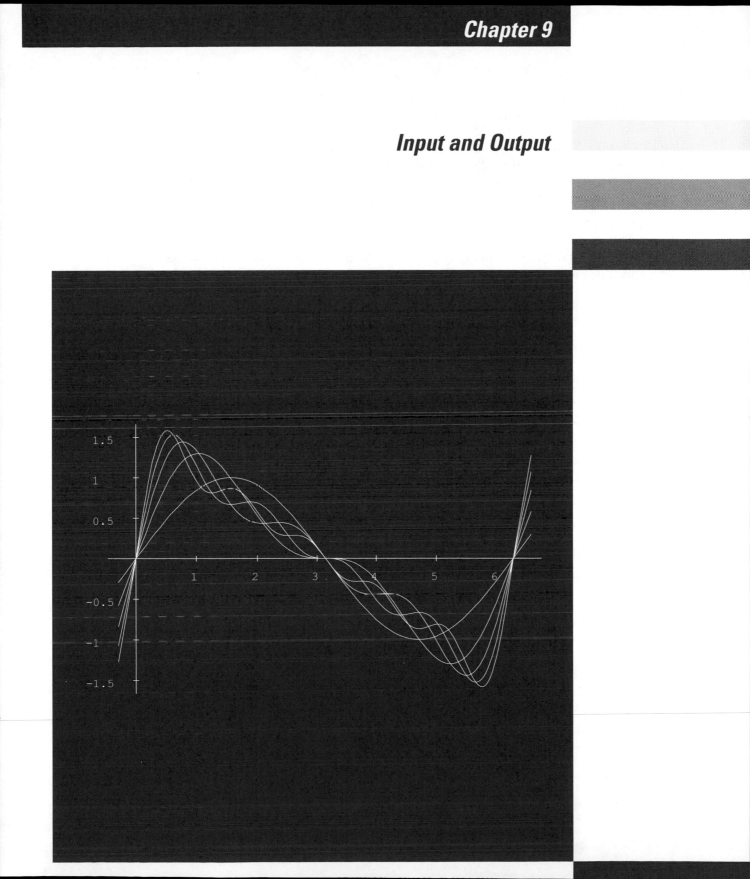

In most programming languages, input and output are among the most tedious and difficult problems. The external form of a value has not much in common with the internal representation. As in graphics, there are many details than can be specified. *Mathematica*, as an interactive language, has certain input and output capabilities built-in. It reads your input and displays it in a two-dimensional form on the screen, taking care of all the formatting. You can also save definitions and results to files without having to be concerned about formatting.

Only if you need to read files that are not written in *Mathematica*'s syntax or want to write files in other than *Mathematica*'s output form do you need to tell the system how to format your output. We can only treat a small number of the possibilities in this book. This, then, is a chapter on selected topics on input and output.

Section 1 is about formatting of output. It tells you how you can change the way *Mathematica* formats expressions.

In section 2 we look at input from files and other programs. This allows you to use *Mathematica* to analyze data obtained from other sources, laboratory measurements or other programs, for example.

Large calculations are the topic of section 3. You can set things up so that *Mathematica* runs unattended, saving its results to a file rather than typing them to the screen.

Finally, we look at some ways of saving all your input and output in a file as you perform interactive calculations. Users of notebooks are fortunate in this respect, the notebook does that automatically.

About the illustration overleaf:

The first five partial sums of the Fourier series of the saw-tooth curve. Including more and more terms gets the approximations closer and closer to the limit curve.

```
l5 = Table[ Sum[Sin[i x]/i, {i, n}], {n, 5} ];
Plot[ Release[l5], {x, -0.3, 2Pi+0.3} ]
```

■ 9.1 Output Formatting

The evaluation of expressions is done without regard to how expressions will eventually appear as output. After the evaluation is complete, the result is formatted for each output file to which it is to be printed. The *format type* of an output file specifies how expressions are to be printed. The most familiar of these format types is `OutputForm`, giving the usual two-dimensional rendering of output.

For each of the format types defined you can give rules that change the way certain expressions are rendered.

■ 9.1.1 Format Types

A format type specifies how expressions are printed. You should be familiar with `OutputForm`, `InputForm`, and `FullForm`. By changing the format type of an output device or file, you can change the way in which expressions are written to that device. When you open a file for writing you can specify the format type with the option `FormatType`. The format type of an already open file, the standard output for example, can be changed with `ResetMedium[]`.

From now on, output will appear in input form.

```
In[1]:= ResetMedium["stdout", FormatType -> InputForm]
```

```
In[2]:= Expand[(a + b)^2]
Out[2]= a^2 + 2*a*b + b^2
```

This is the input form of the output form.

```
In[3]:= OutputForm[%]
                       2           2
Out[3]"//"OutputForm= a  + 2 a b + b
```

Used as a command, a format type formats its argument in the required way, e.g., `FullForm[`*expr*`]`. The result of this formatting is then rendered according to the format type of the output file, as we have just seen in the last example above.

■ 9.1.2 Defining Print Forms

For each format type, you can define rules that change the way a particular expression is printed on an output file that uses that format type. By default, such rules are defined for the format type `OutputForm`.

For example, many mathematical functions have arguments that are normally printed as indices. The function `BesselJ[`*n*`, `*x*`]` is typeset as $J_n(x)$. To have *Mathematica* print an approximation of this, use the definition

```
Format[b:BesselJ[_, _]] := Subscripted[b, 1].
```

Such rules for formatting are stored with their argument and not with the function
Format[].

System symbols for which formats are to be
defined, must be unprotected first.

In[1]:= Unprotect[BesselJ, BesselY];

Since we do not need the names of the
arguments, we only give a name to the whole
pattern.

In[2]:= Format[b:BesselJ[_, _]] := Subscripted[b, 1]

The format is used to print the first argument as
a subscript.

In[3]:= BesselJ[1, x]

Out[3]= BesselJ $_1$[x]

You can also change the head of the expression
to anything else.

In[4]:= Format[BesselY[n_, x_]] := Subscripted[Y[n, x], 1]

BesselY now prints as Y, and with a
subscript.

In[5]:= BesselY[2, z]

Out[5]= Y$_2$[z]

The building blocks you can use to define formats are described in sections 2.6.6
to 2.6.8 of the *Mathematica* book. The formatting command Subscripted[] can take
care of all the simpler cases, in which certain arguments are to be printed as subscripts
or superscripts. In more complicated circumstances, you must assemble the output form
using the primitive operations available.

■ 9.1.3 Application: Tensors

Tensors are a generalization of vectors and matrices used in physics and mathematics.
Here we are only concerned with typesetting problems for tensors. In the usual notation
one might encounter things like $\Gamma_k{}^{ij}(x, y, z, t)$ or $R_{ij}{}^{kl}$. First, we have to design an
input syntax, a way of typing such tensors in *Mathematica*. A possible way is

        Tensor[Gamma][li[k], ui[i], ui[j]][x, y, z, t]

for $\Gamma_k{}^{ij}(x, y, z, t)$ and

        Tensor[R][li[i], li[j], ui[k], ui[l]]

for $R_{ij}{}^{kl}$. We use the two functions ui[] and li[] to denote upper and lower indices.
Some way of indicating which expressions are tensors is necessary. We use Tensor[h]
to denote the tensor h. In order to have these tensors print out in the desired way, we
have to do some programming. (This is one of the more complicated circumstances
mentioned in the previous subsection.)

```
Format[ Tensor[t_][ind___] ] :=
    Block[{indices},
        indices = {ind} /. {ui->Superscript, li->Subscript};
        SequenceForm[t, Sequence @@ indices]
    ]
```

Part of **Tensors.m**: Formatting tensors

The substitution builds a list of subscripts and superscripts. This list is then spliced into the argument list of `SequenceForm[]` which prints its arguments without intervening space.

Here is our first test example.

```
In[1]:= Tensor[Gamma][li[k], ui[i], ui[j]][x, y, z, t]

Out[1]= Gamma  [x, y, z, t]
             k
```
$$\text{Out[1]= Gamma}^{ij}_{\ k} [x, y, z, t]$$

And here is the second one.

```
In[2]:= Tensor[R][li[i], li[j], ui[k], ui[l]]

Out[2]= R
```
$$\text{Out[2]= R}^{kl}_{ij}$$

Rules belong to a certain format type. We have not defined a rule for the format type `TeXForm[]` and therefore no special formatting is done.

```
In[3]:= TeXForm[ Tensor[g][ui[i], li[j]] ]

Out[3]//TeXForm= {\it Tensor}(g)({\it ui}(i),{\it li}(j))
```

Note that we have only defined a rule for objects of the form `Tensor[`*t*`][`*indices*`]`. The first of our examples is of the form `Tensor[`*t*`][`*indices*`][`*arguments*`]`. The *head* of this expression matches the rule for formatting and the elements are then formatted in the default way.

Typesetting is not an easy thing to do, be it with TeX or *Mathematica*. You will probably have to go through a lot of trial and error to get it right.

■ 9.2 Input from Files and Programs

■ 9.2.1 Low-Level Input

The terms "low-level" and "high-level" refer to the amount of details you have to program yourself. In traditional programming languages, files are usually treated at a rather low level. You open a file, perform various read operations, and then close it again. In *Mathematica*, these functions are OpenRead["*file*"], Read["*file*"], and Close["*file*"]. Files are always referred to by their names as strings. There are no "file descriptors" or "logical units". Nevertheless, it is a good idea to store the name of the open file in a variable and use this variable for all future references. A typical program segment that reads *Mathematica* expressions from a file looks like this:

```
ReadLoop[fileName_String] :=
    Block[{file, expr},
        file = OpenRead[fileName];
        If[ Head[file] =!= String, Return[] ];
        While[ True,
            expr = Read[file];
            If[ expr === EndOfFile, Break[] ];
            Print["expr is ", expr]
        ];
        Close[file]
    ]
```

ReadLoop1.m: A simple loop for reading expressions

The command ReadLoop[] receives the name of the file to read as parameter. It then opens that file. OpenRead[] returns the name of the file opened and we assign it to the local variable file. The following While[] loop reads expressions from the open file and prints them out until it encounters the end of the file, indicated by the symbol EndOfFile. Finally, we close the file again.

With external operations there is always a chance for failure. The most common being that the file simply does not exist. In such a case, OpenRead[] prints a message and returns the symbol Null instead of the file name. A good program should test for this. In case of an error, the function returns prematurely. There is no need to print an error message, this has already been done by OpenRead[].

■ 9.2.2 Referring to Open Files

The reason for using the return value of OpenRead[] for all future references to the file, rather than its original name fileName, is that the two could be quite different. Most operating systems provide abbreviated ways of referring to files, and *Mathematica* takes these into account when opening a file. The name returned by OpenRead[] is the fully expanded "absolute" file name. It is under this name that *Mathematica* stores the information associated with all open files it maintains (in the variable $\$\$Media$).

The syntax ~/ refers to a user's home directory in UNIX. *Mathematica* expands this name to the absolute path name shown.

```
In[1]:= file = OpenRead[ "~/init.m" ]
Out[1]= /users/maeder/init.m
```

It does not find the file under its original name.

```
In[2]:= Read[ "~/init.m" ]
General::notopen: ~/init.m is not open.
Out[2]= Read[~/init.m]
```

Always use the name returned by OpenRead[].

```
In[3]:= Read[ file ]
```

At the end, the file should be closed again.

```
In[4]:= Close[ file ]
Out[4]= /users/maeder/init.m
```

■ 9.2.3 Things to Read

Read[*file*] reads a *Mathematica* expression. Read[*file*, *type*] with a second argument can be used to read other things as well. A list of these types is in section 2.6.20 of the *Mathematica* book.

The most interesting feature is the ability to give a *skeleton expression* as the thing to read. All occurrences of the basic types Byte, Number, etc. in this skeleton will be filled from the file and the resulting expression is then returned. The file datafile in the next example contains the numbers 55, 66, 1.11111111, 1, 2, 3, 777 on separate lines to demonstrate some of these ideas.

First, we open the file for reading.

```
In[1]:= file = OpenRead["datafile"]
Out[1]= datafile
```

This reads an integer as a *Mathematica* integer.

```
In[2]:= Read[ file, Number ]
Out[2]= 55
```

Real converts numbers to approximate numbers.

```
In[3]:= Read[ file, Real ]
Out[3]= 66.
```

The number read is inserted in place of the symbol Number in the skeleton given.

```
In[4]:= Read[ file, f[Number] ]
Out[4]= f[1.11111111]
```

Here are three slots to fill and so three numbers are read.

```
In[5]:= Read[ file, {Number, Number, Number} ]
Out[5]= {1, 2, 3}
```

■ 9.2.4 Application: Reading Unevaluated Expressions

When you read an expression, either by Read[*file*] or Read[*file*, Expression], it is evaluated as part of the normal evaluation sequence of *Mathematica*. To return an unevaluated expression, you need to somehow wrap Hold[] or HoldForm[] around it before it gets a chance of being evaluated. The observations made in the previous subsection show a way of doing this. Simply use Read[*file*, Hold[Expression]]. *Mathematica* will fill in an expression from the file in place of the keyword Expression and because this is inside Hold[] it will not be evaluated later on. The following read loop **ReadLoop2.m** reads expressions from a file and prints them out—unevaluated.

```
ReadLoop[fileName_String] :=
    Block[{file, expr},
        file = OpenRead[fileName];
        If[ Head[file] =!= String, Return[] ];
        While[ True,
            expr = Read[file, HoldForm[Expression]];
            If[ expr === EndOfFile, Break[] ];
            Print[ ]; Print["expr is ", expr]
        ];
        Close[file]
    ]
```

ReadLoop2.m: Reading expressions unevaluated

Here is this function applied to the package **ExpandIt.m** from subsection 2.5.2. The individual commands in that file are read as expressions, but they are not evaluated. HoldForm[] is like Hold[], but it is invisible on output.

The value returned by ReadLoop[] is the value of the Close[] command, which returns the name of the file closed.

```
In[1]:= ReadLoop["ExpandIt.m"]

expr is ExpandIt::usage =
  ExpandIt[e] expands all numerators and denominators in\
  e.

expr is Begin['Private']

expr is ExpandIt[x_Plus] := ExpandIt /@ x
```

$$\text{expr is ExpandIt[x\_] := } \frac{\text{Expand[Numerator[x]]}}{\text{Expand[Denominator[x]]}}$$

```
expr is End[]

expr is Null

Out[1]= ExpandIt.m
```

■ 9.2.5 Reading from a Program

Under any reasonable operating system it is possible to read from an external program in the same way you read from a file. Opening the file starts the external program and every read statement will wait for the program to supply enough output to satisfy the request. Closing the file terminates the external program.

Here is a small program that runs the UNIX date command to obtain the current date and time inside *Mathematica*.

```
Date[] :=
    Block[{process, result},
        process = OpenRead["!date"];
        result = Read[process, String];
        If [ result === EndOfFile, Return[] ];
        Close[process];
        result
    ]
```

Date1.m: the first version of the Date[] command

The output of the date command is returned as a string.

```
In[1]:= Date[]
Out[1]- Tue Aug 15 12:28:17 CDT 1989
```

Error checking is different in this case. Opening an external program always succeeds, even if the program does not exist. If the program does not exist, then the first attempt to read from it will return EndOfFile and that is the condition we test.

As a string, the date is not very useful. The output format of the date command is under program control (through command line options) and we can get it to print the date in a form that is acceptable to *Mathematica*, as a list of numbers! We then read it as an Expression instead of a String.

```
Date[] :=
    Block[{process, result},
        process = OpenRead["!date '+{%y, %m, %d, %H, %M, %S}'"];
        result = Read[process, Expression];
        If [ result === EndOfFile, Return[{}] ];
        Close[process];
        result
    ]
```

Date2.m: the second version of the Date[] command

The output of the date command is returned as a list
{*year*, *month*, *day*, *hour*, *minute*, *second*}.

```
In[1]:= Date[]
Out[1]= {89, 8, 15, 12, 28, 19}
```

■ 9.2.6 High-Level Input

High-level input takes care of the details of opening and closing files by itself. The most commonly used input function is undoubtedly Get[*file*], usually used in its prefix form <<*file*.

Another useful command is ReadList[] which is the way to transfer data from other programs or laboratory measurements into *Mathematica*. It reads the whole file and returns a list of all the things read. The things to read are specified in the same way as in Read[] (subsection 9.2.1). In fact, it is an easy exercise to write ReadList[] in terms of Read[]:

```
MyReadList[fileName_String, thing_:Expression] :=
    Block[{file, expr, list = {}},
        file = OpenRead[fileName];
        If[ Head[file] =!= String, Return[list] ];
        While[ True,
            expr = Read[file, thing];
            If[ expr === EndOfFile, Break[] ];
            AppendTo[list, expr]
        ];
        Close[file];
        list
    ]
```

hb MyReadList.m: Writing our own ReadList[] command

It works just like ReadList[], reading all the expressions in the file (they are all numbers) and collecting them in a list.

```
In[2]:= MyReadList[ "datafile" ]
Out[2]= {55, 66, 1.11111111, 1, 2, 3, 777}
```

The contents of the file is read as approximate numbers, that are inserted as arguments of Sin[], which then evaluates to the sine of the numbers in the file.

```
In[3]:= MyReadList[ "datafile", Sin[Real] ]
Out[3]= {-0.999755, -0.0265512, 0.896192, 0.841471,
    0.909297, 0.14112, -0.855551}
```

■ 9.2.7 High-Level Program Input

Again, the file to read can be a program. With the two commands Get["!*program*"] and ReadList["!*program*"] we do not have to start and terminate the program explicitly. Our Date[] function from subsection 9.2.5 now becomes a one-liner:

$$Date[] := << "!date '+{%y, %m, %d, %H, %M, %S}'".$$

■ 9.3 Running *Mathematica* Unattended

Mathematica is quite able to perform calculations that run for hours or even days. If you want to perform a longer computation, then you might want to run it in the background. The following subsections give some hints for doing this on multi-tasking computers. It does not apply to PC-style machines where there is not much point in setting things up differently. You just let *Mathematica* run for as long as you wish.

■ 9.3.1 Batch Mode

This subsection gives some ideas on how to do large calculations in *Mathematica* unattended, often called *batch mode*. Much of this material is specific to the operating system UNIX. It has been tested on Sun-3 and Sun-4 computers under Version 4.0.1 of SunOS using the C-shell. It is trivial to adapt this to other versions of UNIX and it should also be possible to do the same kind of things on other multi-tasking computers.

The main difference between the normal interactive mode of operation and batch mode is that you do not have a chance to react to things that go wrong. All your commands have to be set up in advance in a file from which *Mathematica* will then read its commands. Having solved a smaller problem interactively, you can review your interactive session and assemble the commands in an input file for a later batch run. This file is then given to *Mathematica* as input instead of your keyboard as usual.

The output that normally appears on the screen can be captured in an output file. After the computation is complete, you can look at this file for any error conditions and hopefully find the answer to your problem in it.

The following computation will depend on a parameter n that we can later increase for the larger batch mode calculation.	`In[1]:= n = 3` `Out[1]= 3`
We use the package mentioned in subsection 4.6.4 on page 104.	`In[2]:= << SwinnertonDyer.m`
We compute the nth Swinnerton-Dyer polynomial.	`In[3]:= SwinnertonDyerP[n, x]` `Out[3]= 576 - 960 x^2 + 352 x^4 - 40 x^6 + x^8`
We are interested in how long it takes to prove it irreducible by trying to factor it.	`In[4]:= Timing[ Factor[%] ]` `Out[4]= {0.5 Second, 576 - 960 x^2 + 352 x^4 - 40 x^6 + x^8}`

In order to prepare for a longer computation, with $n = 5$ say, we collect the commands we just entered into a file sw5.in:

```
n = 5
<<SwinnertonDyer.m
SwinnertonDyerP[n, x];
Timing[ Factor[%] ]
```

and then issue the following command at the shell prompt:

```
nice +16 math < sw5.in >& sw5.log &
```

This starts *Mathematica* in the background at a lower priority, connecting the standard input to the file sw5.in and capturing all output in sw5.log. After the computation has finished, the log file will look like this:

```
Mathematica (sun3.68881) 1.2 (May 31, 1989) [With pre-loaded data]
by S. Wolfram, D. Grayson, R. Maeder, H. Cejtin,
   S. Omohundro, D. Ballman and J. Keiper
with I. Rivin and D. Withoff
Copyright 1988,1989 Wolfram Research Inc.

In[1]:=
Out[1]= 5

In[2]:=
In[3]:=
In[4]:=
Out[4]= {400.583 Second, 2000989041197056 - 44660812492570624 x  +

                          4                        6                       8
>      183876928237731840 x  - 255690851718529024 x  + 172580952324702208 x  -

                           10                      12                      14
>      65892492886671360 x   + 15459151516270592 x   - 2349014746136576 x   +

                         16                    18                   20
>      239210760462336 x   - 16665641517056 x   + 801918722048 x   -

                      22                 24               26           28          30
>      26625650688 x   + 602397952 x   - 9028096 x   + 84864 x  - 448 x  +

           32
>      x  }

In[5]:=
```

You notice that the input lines are empty. Normally the operating system, and not *Mathematica*, echoes the input back to the terminal. Here you want *Mathematica* itself to echo all input to the standard output. The variable $Echo is a list of files to which input is echoed. It is normally empty. To have your input appear in the output file, put the following *Mathematica* command at the beginning of your input file:

$$\text{AppendTo[\$Echo, "stdout"]; \$Line = 0; .}$$

This also resets the line numbers so that the rest of the calculations starts with input line 1 as before (this is optional, of course).

It might be useful to have *Mathematica* print its output in `InputForm` so it could be input into another computation easily. You can either set `$PrePrint = InputForm` (see subsection 8.2.1), or you can change the format type of the standard output like this:

`ResetMedium["stdout", FormatType->InputForm].`

■ 9.3.2 Infinite Calculations

An infinite calculation is one that you can run for as long as you wish (or until the computer crashes) and that produces one result after another. You can use this to try to find counterexamples to a conjecture or to find numbers with a certain property by starting with $n = 1$ and testing for $n = 1, 2, 3, \ldots$ for as long as your patience lasts.

In *Mathematica* you write a short program that contains an infinite loop and repeatedly does some computation. Since it potentially never returns (you have to interrupt the computation to get back at the next input prompt), you should use `Print[]` statements inside the loop to inform yourself about the progress of the computation. For an example, let us look at the famous $3n+1$-problem, also called the Collatz problem. Starting with an integer k_1, we construct a sequence of integers $k_1, k_2, k_3, \ldots$ according to the following formula:

$$k_{i+1} = \begin{cases} k_i/2 & k_i \text{ even} \\ 3k_i + 1 & k_i \text{ odd}. \end{cases}$$

If you try this out, you will notice that after a while one of the k_i becomes 1 and then the sequence repeats itself with 1, 4, 2, 1, ... We want to find the integer with the longest sequence before hitting 1. This function computes this length:

```
length[n_Integer?Positive] :=
    Block[{i=1, m=n}, While[m != 1, m = If[OddQ[m], 3m+1, m/2]; i++];i ]
length[_] = 0
```

The length of the Collatz sequence

Now we write our infinite loop. It computes `length[i]` for i from some lower bound on upwards. Whenever it finds a new maximum, it prints a line. The only way to stop it is to interrupt *Mathematica* or to kill the process.

```
max[low_] :=
    Block[{m=0, n=0, i=low, j},
        While[ True,
            j = length[i];
            If[j > m, m = j; n = i; Print["length[", n, "] = ", m] ];
            i++
        ]
    ]
```

Part of **Collatz.m**: Finding the maximum

Starting with 27, it takes 112 steps before reaching 1.

```
In[2]:= length[27]

Out[2]= 112
```

We start the computation at 1 and let it run for a while before interrupting it.

```
In[3]:= max[1]

length[1] = 1
length[2] = 2
length[3] = 8
length[6] = 9
length[7] = 17
length[9] = 20
length[18] = 21
length[25] = 24
length[27] = 112
length[54] = 113
length[73] = 116
length[97] = 119
length[129] = 122
length[171] = 125
```

This kind of computation is of course very well suited for running in batch mode, as explained in the previous subsection. The following command file collatz.in could be used:

```
AppendTo[ $Echo, "stdout" ]; $Line = 0;
<< Collatz.m
max[1]
```

Performing the computation in the background and redirecting its output to the file collatz.log, you could then examine the output file from time to time to see how far the computation has progressed.

Apart from printing the maxima as they are found, it is also a good idea to print some information about the progress of the computation. Should the computer crash while you are running your computation, you would not know where to restart it and would have to go back to the last maximum found. The following variant of max[] prints the value of the loop variable after every 100 iterations.

```
max[low_] :=
    Block[{m=0, n=0, i=low, j},
        While[ True,
            j = length[i];
            If[j > m, m = j; n = i; Print["length[", n, "] = ", m] ];
            i++;
            If[ Mod[i, 100] == 0, Print["i = ", i] ]
        ]; Null
    ]
```

Keeping informed about the progress

Should the computation abort abnormally, you can simply restart it at the last value reported. This is the reason we provided for the argument `low` in the definition of `max[]`.

If you really want to hunt for new records then it pays to look more closely at the definition of our sequence. It does not make much sense, for example, to check the length for an even integer since the first thing we do is divide it by two. We could therefore always start with an odd number and increase the loop variable `i` by 2 in each iteration. The `length[]` function is indeed rich in interesting patterns. Some of these can be revealed by a simple plot of its values for the first few hundred integers.

We compute the first 300 values of the length function.

`In[4]:= vals = Array[length, 300];`

And then we draw a dot for each of its values.

`In[5]:= ListPlot[ vals, PlotRange->All ]`

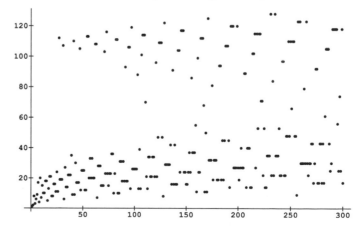

■ 9.4 Session Logging

This section shows some ways of recording the input or output of a *Mathematica* session in a file. If you use *Mathematica* with the notebooks frontend, all input and output is saved in a notebook and you do not have to use any of the mechanisms explained in this section.

■ 9.4.1 Keeping a Log of the Input

The standard package **Utiltities/Record.m** contains the single command

```
AppendTo[ $Echo, OpenAppend["math.record"] ]
```

that will cause all input to be written to the file math.record.

An output channel like $Echo is a list of files to which certain output is written. A list of all such channels is in section 2.6.16 of the *Mathematica* book. By default, $Echo is the empty list and *Mathematica* does not echo its input, as we have seen in subsection 9.3.1.

■ 9.4.2 Logging Input and Output

If you want to keep a record of *Mathematica*'s output, you can similarly use

```
AppendTo[ $Output, OpenWrite["math.log"] ].
```

The default format type for a newly opened file is InputForm so the record will be in a form that is easy to read back into *Mathematica* later on.

$Output is of course not empty to begin with. Output is now written to two files, one being the standard output, the other one the just-opened math.log.

```
In[1]:= AppendTo[ $Output, OpenWrite["math.log"] ]
Out[1]= {stdout, math.log}

In[2]:= Expand[(a+b)^2]
           2
Out[2]= a  + 2 a b + b
                        2
```

After this session, math.log contains the following text:

```
In[2]:=
Out[2]= a^2 + 2*a*b + b^2

In[3]:=
```

We notice that it also contains the input and output prompts (these prompts are printed just like other output and therefore they appear in all files of the channel $Output). Perhaps we can put both the input and output into the same file?

```
OpenLog::usage = "OpenLog[filename] starts logging all input and output to filename."
CloseLog::usage = "CloseLog[] closes the logfile opened by OpenLog[]."

Begin["`Private`"]

logfile=""

OpenLog[filename_String] := (
    logfile = OpenWrite[filename];
    If[ Head[logfile] =!= String, Return[] ];
    AppendTo[$Echo, logfile];
    AppendTo[$Output, logfile];
    )

CloseLog[] := (
    $Echo = Complement[$Echo, {logfile}];
    $Output = Complement[$Output, {logfile}];
    Close[logfile];
    )

End[]
Null
```

SessionLog.m: Logging input and output

OpenLog[] opens the file given as argument and, if it succeeds, appends it to both
$Echo and $Output. CloseLog[] removes the log file from both $Echo and $Output
and then closes it. If you do not close the log file before the end of the session, it will
be closed by *Mathematica* as part of the exit processing.

We want all input and output to go to the file session.log.	`In[2]:= OpenLog["session.log"]`
Here we generate some output.	`In[3]:= Expand[(x+y)^3]`
	$Out[3]= x^3 + 3 x^2 y + 3 x y^2 + y^3$
And then close the log file.	`In[4]:= CloseLog[]`

After the above session, session.log contains the following text:

```
In[3]:= Expand[(x+y)^3]
Out[3]= x^3 + 3*x^2*y + 3*x*y^2 + y^3
In[4]:= CloseLog[]
```

If you want the output to be rendered in output form, use

```
OpenWrite[filename, FormatType->OutputForm]
```

to open the log file in the command `OpenLog[]`.

■ 9.4.3 Advanced Topic: A Sophisticated Transcript

The logging mechanism from the previous subsection prints input and output lines the way they appear in the session, including input and output prompts. Now we want to solve the following problem:

- The log should contain the input line number and the input enclosed in a comment.
- The output should be placed in the file without the output prompts. It should be in input form, unaffected by any special formats.

The second requirement means that if we type `Short[`*expr*`]`, then we want *expr* itself in the log file and not the short form. What we want is, in fact, the value assigned to the corresponding `Out[`*n*`]`, which does this kind of transformation (see section 2.6.4 of the *Mathematica* book for an explanation).

In a similar way we can get the input line from the value of `In[`*n*`]`. In both cases, we must not evaluate the values. It is quite tricky to get the input form of an unevaluated expression as a string. The auxiliary function `GetInputForm[]` does it. It takes a `ValueList` as argument. A value list is the object returned by `DownValue[`*sym*`]`, a function that returns all rules defined for *sym*. In our case *sym* is `In` and `Out`.

The command that writes the input and output values to the log file is assigned to `$Post`. It is therefore called once for each evaluation. Since the current output value has not been assigned to `Out[`*n*`]` at the time `$Post` is evaluated, we write out the input and output from the *previous* command. This makes it necessary to delay the start of the logging process by one and to call it an extra time when we want to close the log file with the function `EndTranscript[]`.

```
Transcript::usage = "Transcript[filename] opens filename for a transcript
    of the current session. It clobbers $Post"
EndTranscript::usage = "EndTranscript[] closes the transcript opened with Transcript[]."

Begin["`Private`"]

`transcript=""

Transcript[filename_String] := (
    transcript = OpenWrite[filename];
    If[ Head[transcript] =!= String, Return[transcript] ];
    $Post := ($Post := MakeTranscript; Identity);
    transcript
    )

EndTranscript[] := (
    MakeTranscript;
    $Post =. ;
    Close[transcript]
    )

GetInputForm[v_ValueList] :=
    Block[{input},
        input = MapAt[HoldForm, v, {1, 2}] [[1,2]];
        input = Characters[ToString[InputForm[input]]];
        input = Drop[input, {1, 9}]; (* HoldForm[ *)
        input = Drop[input, -1];     (* ]        *)
        StringJoin @@ input
    ]

MakeTranscript :=
    Block[{input, output},
        input  = GetInputForm[Take[DownValue[In],  {-2}]];
        output = GetInputForm[Take[DownValue[Out], {-1}]];
        WriteString[transcript, "(* In[", $Line-1, "]=\n", input, "\n*)\n"];
        WriteString[transcript, "\n", output, "\n\n"];
        Identity
    ]

End[]
Null
```

Transcript.m: A sophisticated logging function

This starts the logging process.

```
In[1]:= Transcript["math.script"]

Out[1]= math.script
```

As usual, we generate some sample output.

```
In[2]:= Nest[1 + 1/#&, x, 4]
```

$$Out[2]= 1 + \cfrac{1}{1 + \cfrac{1}{1 + \cfrac{1}{1 + \cfrac{1}{x}}}}$$

In the output, we see the matrix form of the result.

```
In[3]:= MatrixForm[IdentityMatrix[4]]

                    1  0  0  0

                    0  1  0  0

                    0  0  1  0

Out[3]//MatrixForm= 0  0  0  1
```

This will do for now.

```
In[4]:= EndTranscript[]

Out[4]= math.script
```

Now let us look at the transcript generated.

```
(* In[1]=
Transcript["math.script"]
*)

"math.script"

(* In[2]=
Nest[1 + 1/#1 & , x, 4]
*)

1 + (1 + (1 + (1 + x^(-1))^(-1))^(-1))^(-1)

(* In[3]=
MatrixForm[IdentityMatrix[4]]
*)

{{1, 0, 0, 0}, {0, 1, 0, 0}, {0, 0, 1, 0}, {0, 0, 0, 1}}
```

Note especially that in the last evaluation the *input form* of the result is given.

If you want to save all the values assigned to Out[] during the session so far you could also simply give the command Save["*file*", Out]. To do this automatically before exiting *Mathematica*, you can put

```
$Epilog := Save["session.m", Out]
```

into your **init.m** file.

Chapter 10

Notebooks

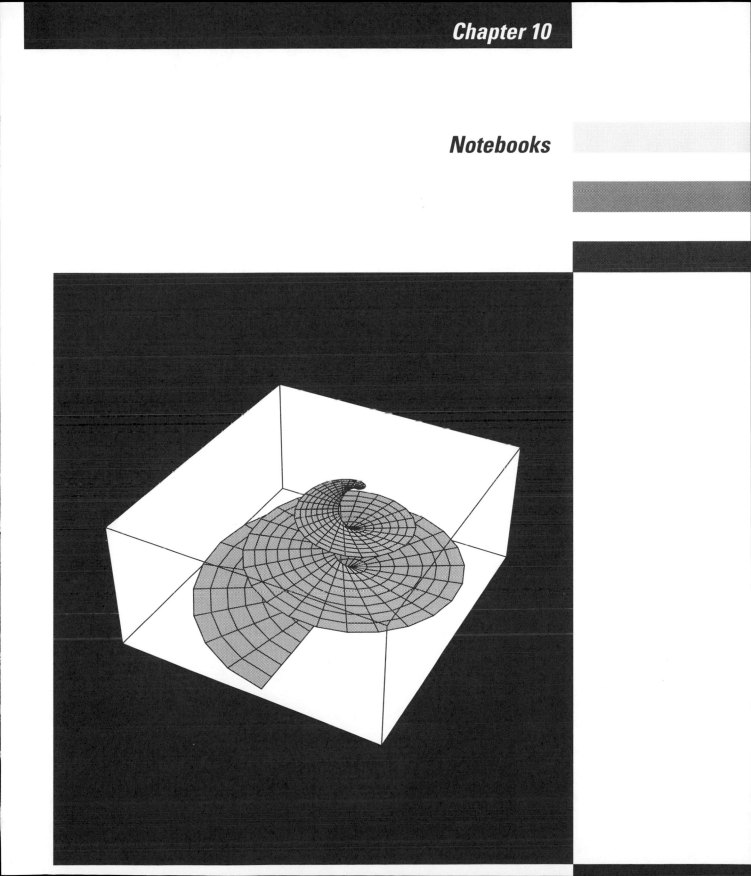

This chapter treats some of the issues concerning the use of *Mathematica* on computers with or without the notebook frontend. It is most useful, if you work on both of these. A package developed under the guidelines in this book can be used on any version of *Mathematica*, since it does not use any of the frontend specific features. On the other hand, if you develop notebooks it takes some effort to convert them into plain packages.

Section 1 compares notebooks and packages and mentions a few things to keep in mind when working on different systems. It also suggests a way of developing packages as notebooks.

Section 2 is about converting packages to notebooks and notebooks to packages. We look into how a notebook is stored as a file.

About the illustration overleaf:

A "spiral" spiral staircase.

```
ParametricPlot3D[{r(1+phi/2)Cos[phi], r(1+phi/2)Sin[phi], -phi/2},
    {r, 0.1, 1.1, 0.125}, {phi, 0, 11Pi/2, Pi/12}]
```

■ 10.1 Packages and Notebooks

Notebooks are structured documents that can contain *Mathematica* input and output, graphics, and ordinary text. You can structure the information in the same way as it is structured in a book, defining chapters, sections, subsections, and so on. Notebooks are a feature of *Mathematica frontends*, programs that provide a sophisticated interface between the user and the *kernel*, that does the computations. These frontends are not available for all versions of *Mathematica*. Without a special frontend, you work with *Mathematica* by typing input at the In[*n*]:= prompt and the results are displayed at the next Out[*n*] = prompt. This is the way of interaction used for the "live" calculations throughout this book. On some machines, graphics output appears in separate windows.

■ 10.1.1 Notebooks as Packages

A typical notebook consists of some definitions and then examples of their use. If you worked mostly with notebooks, you probably did not bother to set up package contexts for them. If you think about using the definitions in your notebooks in other places as well, you should design them in the same way as you design packages.

The package part of a notebook is everything in the initialization cells. These initialization cells can be grouped together and are normally given the heading "Implementation". In that implementation part you can use the same framework for setting up the contexts as you do in a package. The commands BeginPackage[], Begin[], EndPackage[], and End[] should be put into separate cells. In the examples in this book, we have used blank lines to separate the parts of a package. These parts are best put into separate cells. Comments can go into text cells instead of using the notation (* *comment* *).

If you need the definitions from a notebook in another notebook, you can simply open the notebook and execute the initialization cells. The definitions are then made in the *Mathematica* kernel and are available in all other notebooks that are open since the kernel is shared among all open notebooks. If you use a local kernel (on the same computer as the front end), you can also read in the notebook into another notebook with the command <<*Notebook*.m. The only things that will be interpreted by the kernel are the initialization cells. Everything else is treated as comment by the kernel. This is obviously not possible if the kernel runs on a different computer that has no access to the locally stored files.

■ 10.1.2 Notebooks that Depend on Packages

You specify that a notebook needs a certain package in the same way as you do in other packages. Either put the command <<*Package*.m in your notebook or better yet Needs["*Package*`"]. This command is best put into an initialization cell. It will then be executed whenever you open the notebook. Some computers have rather strange ways of referring to files on their hard disk. Using `Needs["`*Package*`'"]` avoids all those problems since *Mathematica* knows how to convert a context name into a file name.

■ 10.1.3 Developing Packages as Notebooks

The way you best develop a new package is different on computers with the notebook frontend available. On workstations without the frontend, you typically keep the package you are working on in an editor, make changes and then read it back into the *Mathematica* session. We have given some hints on how to set things up during the debugging phase in subsection 2.3.1 on page 43.

In a notebook frontend, you might want to set things up differently. (And only because certain computers still do not support multitasking.) You should still insert the statement `Clear[`*syms*`...]` near the beginning of the notebook under development and make the cells containing the context-switching statements *inactive*. Instead of reading in the new version after making some corrections, you simply select all the cells in the notebook that are part of the package and evaluate them. You can keep the test examples for your code in the same notebook and then evaluate them again to see if the changes have fixed the problems.

■ 10.2 Conversion between Notebooks and Packages

■ 10.2.1 Packages to Notebooks

If you use *Mathematica* on a computer with the notebooks interface and receive a file containing a package that you want to use, you have to convert it to a notebook. The first time you open a file that is not a notebook (sometimes called a plain ASCII file) inside the frontend, it will be converted automatically. There are certain choices on how this conversion should take place.

For packages, it is best to ask that every blank line become a cell boundary. We have used meaningful blank lines in all the examples presented in this book. After reading in the package (now turned notebook), you should do some editing to make it look nice. First, all cells containing *Mathematica* input should be made initialization cells. Then you might want to group cells, insert text cells for comments, and close the subgroups.

■ 10.2.2 Notebooks to Packages

Treating a notebook as a package is easy: It already *is* a package. Everything but the initialization cells is ignored by *Mathematica* when you read it in with Get[].

Nevertheless, if you want to distribute a notebook as a package, you should do some editing. Notebooks contain a lot of information that is of no use on a computer without the notebook frontend. Notebooks contain the unformatted output of all calculations. In the case of graphics this means the whole POSTSCRIPT code! It is therefore a good idea to delete graphics cells from the notebook before saving it for transfer to another computer that does not have a frontend. Frontends can strip a lot of this useless text if you choose the appropriate options for saving the notebook.

To make the resulting file more readable, you should also remove the special codes that describe its cell structure and put text cells into comments instead. You can also request that a blank line be put between cells.

■ 10.2.3 Example: Skeleton.m

The package **Skeleton.m** (see section 2.4) was converted into a notebook and then saved again as an ordinary ASCII file. The result is reproduced in part below.

When **Skeleton.m** was opened as a notebook, the options for opening files were set so as to request a new cell for each blank line. After that we made a couple of modifications inside the frontend.

- Comments appearing on a line by themselves were converted to text cells.
- All remaining cells were made initialization cells.

- We added a new cell containing the title "Skeleton.m" and a cell with the subtitle "Implementation".
- The text cells and the initialization cells following them were grouped together and the groups were closed.

The resulting notebook was then saved and transferred back to the computer used for typesetting this book for inclusion in here.

```
(*^

::[paletteColors = 128;
    fontset = title, "Chicago", 24, L3, center, bold, nohscroll;
    fontset = subtitle, "Chicago", 18, L2, center, bold, nohscroll;
        ⋮
    fontset = special5, "Chicago", 12, L2;]
:[inactive; font = title; ]
Skeleton.m
:[inactive; startGroup; Cclosed; font = subtitle; ]
Implementation
:[inactive; startGroup; Cclosed; font = text; ]
set up the package context, included any imports
:[initialization; font = input; ]
*)
BeginPackage["Skeleton`", "Package1`", "Package2`"]
(*
:[initialization; endGroup; font = input; ]
*)
Needs["Package3`"]    (* read in any hidden imports *)
(*
:[inactive; startGroup; Cclosed; font = text; ]

usage messages for the exported functions and the context itself
:[initialization; font = input; ]
*)
Skeleton::usage = "Skeleton.m is a package that does nothing."
    ⋮
(*
:[inactive; startGroup; Cclosed; font = text; ]
set the private context
:[initialization; endGroup; font = input; ]
*)
Begin["`Private`"]
    ⋮
(*
:[inactive; startGroup; Cclosed; font = text; ]

definition of the exported functions
```

```
:[initialization; font = input; ]
*)
Function1[n_] := n
    ⋮
(*
:[inactive; startGroup; Cclosed; font = text; ]
Epilogue
:[initialization; font = input; ]
*)
End[]          (* end the private context *)
(*
:[initialization; font = input; ]
*)
Protect[ Function1, Function2 ]    (* protect exported symbols *)
(*
:[initialization; endGroup; endGroup; font = input; ]
*)
EndPackage[]  (* end the package context *)

(*
^*)
```

Skeleton.m converted to a notebook

Such a file is rather difficult to read for a human, but *Mathematica* will have no problem. Everything but the commands from the original package is inside comments (* ... *). The lines that look like this

$$:[cell type;\ attribute;\ldots;\ ]$$

define the beginning of a new cell. If you are familiar with notebooks, you should recognize most of the attributes appearing on these lines.

Chapter 11

Producing the Cover Picture

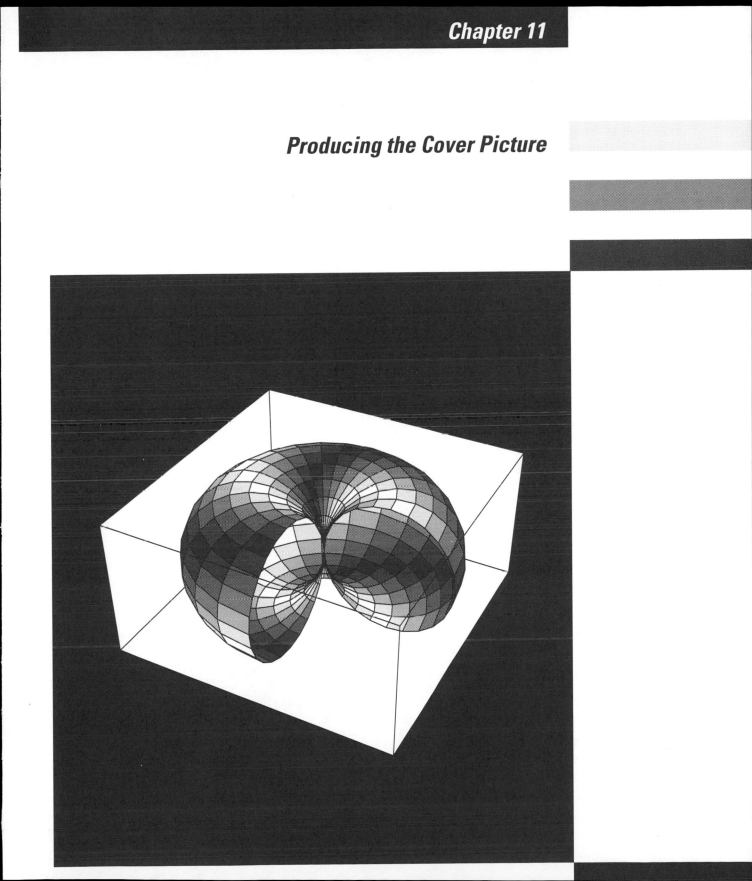

This chapter explains in detail how the cover picture of the book was produced. For this picture, we needed an extension of the standard package **ParametricPlot3D.m** that is explained in section 1. This section also contains some utilities for coloring plots that should be of general interest.

In section 2 we develop the cover picture step by step.

About the illustration overleaf:

This shape gave me the inspiration for the cover picture. For this chapter, it is shaded in a diagonal pattern, different on the inside and the outside.

```
SphericalPlot3D[ {Sin[theta],
    FaceForm[GrayLevel[0.05 + 0.9 Sin[2theta + phi]^2],
             GrayLevel[0.05 + 0.9 Sin[2theta - phi]^2]]},
    {theta, 0, Pi, Pi/24}, {phi, 0, 3Pi/2, Pi/12}, Lighting->False ]
```

■ 11.1 Extensions to Graphics Packages

■ 11.1.1 A New ParametricPlot3D

The command `Plot3D[]` allows you to define the color or gray level of the surface plotted under program control. The form is `Plot3D[{z, *style*},...]`, where *style* can be `RGBColor[]` or `GrayLevel[]`, for example. The cover of the *Mathematica* book was produced in this way. We can do the same thing for parametric plots. Normally a parametric plot is given as

$$\text{ParametricPlot3D}[\{x, y, z\}, \{u, u0, u1\}, \{v, v0, v1\},...]$$

with *x*, *y*, and *z* being functions of *u* and *v*. If we want to specify the color of the surface patches as a function of *u* and *v*, a natural extension of the above syntax is

$$\text{ParametricPlot3D}[\{x, y, z, \textit{style}\}, \{u, u0, u1\}, \{v, v0, v1\},...].$$

We can do this by writing a different version of the auxiliary function `MakePolygons[]` in **NewParametricPlot3D.m**. We looked at this function in subsection 4.6.2. The new version is below.

```
MakePolygons[vl_List] :=
    Block[{l, ll, mesh, cols},
        l = Map[Take[#, 3]&, vl, {2}]; (* the coords *)
        ll = Map[RotateLeft, l];
        cols = Map[#[[4]]&, vl, {2}];  (* the colors *)
        mesh = {l, RotateLeft[l], RotateLeft[ll], ll};
        mesh = Map[Drop[#, -1]&, mesh, {1}];
        mesh = Map[Drop[#, -1]&, mesh, {2}];
        cols = Drop[cols, -1];
        cols = Map[Drop[#, -1]&, cols, {1}];
        mesh = Polygon /@ Transpose[ Map[Flatten[#, 1]&, mesh] ];
        Flatten[Transpose[{Flatten[cols], mesh}]]
    ] /; TensorRank[vl] == 3 && Dimensions[vl][[3]] == 4
```

A second rule for `MakePolygons[]` in **NewParametricPlot3D.m**

When `MakePolygons[]` is called, `vl` is a matrix of coordinate triplets or quadruplets in this case. We first extract the first three elements from each of the entries of this matrix and perform the same computations with them as we did in the old case. Then we extract the style commands that are in the fourth elements of the entries and rearrange them in the same way as we do the polygons. Finally, the polygons and the styles are combined in a way that each style element comes just before the corresponding polygon. This polygon will therefore be drawn in the required style.

This new version of ParametricPlot3D is in the file **NewParametricPlot3D.m**.

```
In[1]:= << NewParametricPlot3D.m
```

In order to see the colors or gray levels we have to turn off lighting of the surface.

```
In[2]:= ParametricPlot3D[ {u - v, u + v, u v,
            GrayLevel[ArcTan[u, v]/2/Pi + 0.5]},
         {u, -1, 1}, {v, -1, 1}, Lighting->False,
            ViewPoint->{-2.4, -1.3, 1.8} ]
```

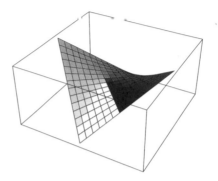

■ 11.1.2 Plot Utilities

The commands RGBColor[] and GrayLevel[] are rather finicky about their arguments. They do not like numbers outside of the interval {0, 1}. We want to define some utilities for specifying colors and gray levels that are easier to use.

The first set of commands transforms the phase angle or argument of a complex number into a gray level or color value. Since the phase angle goes from $-\pi$ to π (once around the circle) it is very natural to let it determine the hue of a color. The function HSBColor[] in the standard package **Graphics/Colors.m** can be used to translate hues into RGB values. Since the arguments of HSBColor[] have to lie in the interval {0, 1} as well, we have to squeeze the value from $-\pi$ to π into that range. This is done in the function ArgColor[]. ArgShade[] uses gray levels instead of colors.

```
ArgColor[z_] := RGBColor[1,1,1]    /; z == 0.0
ArgColor[z_] := HSBColor[ N[(Pi + Arg[z])/(2Pi)], 1, 1 ]

ArgShade[z_] := GrayLevel[1]    /; z == 0.0
ArgShade[z_] := GrayLevel[N[(Pi + Arg[z])/(2Pi)] ]
```

Mapping phase angles to colors and gray levels

The reason for the special rule when z is equal to 0.0 is that for 0 the argument is undefined and the function Arg[z] therefore does not evaluate to a number.

For the cover picture, we used the function `ColorCircle[]`, that translates its argument into a hue value and takes it modulo 2π. It is similar to code found inside `HSBColor[]`. The optional second argument specifies the overall brightness of the colors.

```
ColorCircle[arg_?NumberQ, light_:1.0] :=
    Block[{ mh = 3.0 Mod[N[arg / Pi], 2.0], ind, frac, frac1, scale },
        ind = Floor[mh]; frac = mh - ind;
        scale = Max[ Min[light, 1.0], 0.0 ];
        frac1 = scale(1.0-frac); frac *= scale;
        Switch[ind,
            0,    RGBColor[scale, frac,   0.0],
            1,    RGBColor[frac1, scale,  0.0],
            2,    RGBColor[0.0,   scale,  frac],
            3,    RGBColor[0.0,   frac1,  scale],
            4,    RGBColor[frac,  0.0,    scale],
            5,    RGBColor[scale, 0.0,    frac1],
            6,    RGBColor[scale, frac,   0.0]
        ]
    ]

ColorCircle[_, light_:1.0] := RGBColor[light, light, light]
```

Mapping to the color circle

The function value is the absolute value of the sin function; the gray level is determined by the phase angle of the sin function. Using `ArgColor[]` on a color display instead of `ArgShade[]`, the sharp jump from black to white does not occur because the colors close up nicely on the color circle.

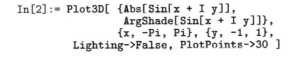

```
In[2]:= Plot3D[ {Abs[Sin[x + I y]],
                 ArgShade[Sin[x + I y]]},
               {x, -Pi, Pi}, {y, -1, 1},
               Lighting->False, PlotPoints->30 ]
```

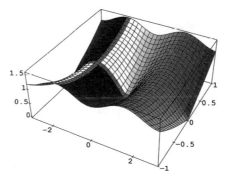

As an aside, note that in the last example the `Sin[]` function is computed *twice* for each point. You can save much time by using the following command instead:

```
Plot3D[ {Abs[z = Sin[x + I y]], ArgShade[z]},...].
```

Here is the complete package **PlotUtilities.m**.

```
BeginPackage["PlotUtilities`", "Graphics`Colors`"]

ArgColor::usage = "ArgColor[z] gives a color value whose hue is proportional
    to the argument of the complex number z."
ArgShade::usage = "ArgShade[z] gives a gray level proportional
    to the argument of the complex number z."

ColorCircle::usage = "ColorCircle[r, (light:1)] gives a color value whose hue
    is proportional to r (mod 2Pi) with lightness light."

Begin["`Private`"]

ArgColor[z_] := RGBColor[1,1,1]    /; z == 0.0
ArgColor[z_] := HSBColor[ N[(Pi + Arg[z])/(2Pi)], 1, 1 ]

ArgShade[z_] := GrayLevel[1]    /; z == 0.0
ArgShade[z_] := GrayLevel[N[(Pi + Arg[z])/(2Pi)] ]

ColorCircle[arg_, light_:1.0] :=
    Block[{ mh = 3.0 Mod[N[arg / Pi], 2.0], ind, frac, frac1, scale },
        ind = Floor[mh]; frac = mh - ind;
        scale = Max[ Min[light, 1.0], 0.0 ];
        frac1 = scale(1.0-frac); frac *= scale;
        Switch[ind,
            0,    RGBColor[scale, frac,    0.0],
            1,    RGBColor[frac1, scale,   0.0],
            2,    RGBColor[0.0,   scale,   frac],
            3,    RGBColor[0.0,   frac1,   scale],
            4,    RGBColor[frac,  0.0,     scale],
            5,    RGBColor[scale, 0.0,     frac1],
            6,    RGBColor[scale, frac,    0.0]
        ]
    ] /; NumberQ[N[arg]]

ColorCircle[_, light_:1.0] := RGBColor[light, light, light]

End[]

Protect[ArgColor, ArgShade, ColorCircle]

EndPackage[]
```

PlotUtilities.m

■ 11.2 The Cover Picture

■ 11.2.1 The Basic Idea

The picture on the cover is a variant of a simple spherical plot. A spherical plot of $\sin\theta$
gives a doughnut whose central hole has shrunk to a single point.

These are the packages that we need.

```
In[1]:= << NewParametricPlot3D.m; << PlotUtilities.m
```

A special kind of doughnut.

```
In[2]:= SphericalPlot3D[ Sin[theta],
          {theta, 0, Pi, Pi/24}, {phi, 0, 3Pi/2, Pi/12},
          Boxed->False ]
```

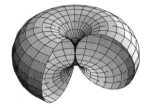

■ 11.2.2 Changing the Radius

The next idea is to vary the radius of the figure as a function of ϕ. The formula
$r = 2 + \cos\phi$ gives a radius that lies between 1 and 3, and closes up after one revolution.

Changing the radius.

```
In[3]:= SphericalPlot3D[ Sin[theta](2 + Cos[phi]),
          {theta, 0, Pi, Pi/24}, {phi, 0, 2Pi, Pi/12},
          Boxed->False ]
```

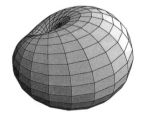

With the formula $r = 2 + \cos(\phi/2)$, it takes two revolutions to close the picture. The surface will self-intersect and in order to see the interior we draw only the bottom half.

```
In[4]:= SphericalPlot3D[ Sin[theta](2 + Cos[phi/2]),
          {theta, Pi/2, Pi, Pi/24}, {phi, 0, 4Pi, Pi/12},
          Boxed->False ]
```

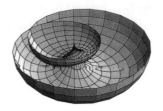

■ 11.2.3 Coloring the Surface

We want the color of the surface to depend on ϕ. For a color display, we can use `ColorCircle[phi]`, but for the pictures in this section we use `GrayLevel[phi/2/Pi]`. When ϕ goes from 0 to 2π, the argument `phi/2/Pi` of `GrayLevel[]` goes from 0 to 1, or from black to white.

Here we use the alternate syntax of `SphericalPlot3D[]` that was developed in section 11.1.1. It allows us to specify the color of the surface at each point.

```
In[5]:= SphericalPlot3D[ {Sin[theta],
                          GrayLevel[phi/2/Pi]},
          {theta, 0, Pi, Pi/24}, {phi, 0, 2Pi, Pi/12},
          Boxed->False, Lighting->False ]
```

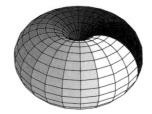

Using `FaceForm[]`, we can specify different colors for the front and back of the polygons.

```
In[6]:= SphericalPlot3D[ {Sin[theta],
             FaceForm[GrayLevel[phi/2/Pi + 1/8],
                 GrayLevel[7/8 - phi/2/Pi]]},
          {theta, 0, Pi, Pi/24}, {phi, 0, 3Pi/2, Pi/12},
          Boxed->False, Lighting->False ]
```

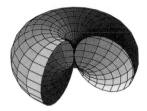

■ 11.2.4 Putting it All Together

The final picture is composed of four segments, two of them showing the whole surface for $0 < \theta < \pi$ and the two smaller ones showing only the lower half, $\pi/2 < \theta < \pi$. In order for the four parts to fit together properly, we fix the plot range to the same value for all four partial pictures. The redefinition of the `DisplayFunction` option prevents the partial pictures from being drawn. They are merely generated and then combined. Here is the complete code that will produce the picture. It takes a long time to compute.

```
Needs["ParametricPlot3D`", "NewParametricPlot3D.m"]
Needs["PlotUtilities`"]

t0 = 0.001; t1 = N[ Pi - t0 ]
dt = (t1 - t0)/36; dp = N[ Pi/20 ]
plotrange = {{-2, 3}, {-2.5, 2.5}, {-1.5, 1.5}}

part[t0_, phi0_, phi1_] := Block[{theta, phi},
    SphericalPlot3D[{Sin[theta] (2+Cos[phi/2]),
                   FaceForm[ColorCircle[phi/2, 1], ColorCircle[phi/2, 0.7]]},
                  {theta, t0, t1, dt}, {phi, phi0, phi1, dp},
                  PlotRange -> plotrange, DisplayFunction -> Identity]
    ]

coverpage := Block[{glist},
    glist = {part[t0, 0, 3Pi/2],     part[Pi/2, 3Pi/2, 4Pi/2],
             part[t0, 4Pi/2, 7Pi/2], part[Pi/2, 7Pi/2, 8Pi/2]};
    Show[ glist, Boxed -> False, RenderAll -> True, Lighting -> False,
         DisplayFunction -> $DisplayFunction ]
    ]
```

CoverPage.m: Drawing the cover of this book

Exercises

This book should teach you enough about *Mathematica* to use it to help you solve the problems in your own area of research or teaching. While the examples chosen always belong to a specific discipline of science or mathematics, you can use the same *methods* for your own purposes. The best exercise you can do to test your understanding of *Mathematica*'s programming language is to try to write a package that implements some of the algorithms you use in your own work.

Nevertheless, this appendix contains a few exercises related to the material covered in this book followed by sketchy solutions.

About the illustration overleaf:

A random walk. We start at the origin and then randomly choose a direction to follow for a line of length 1. This picture consists of 5000 segments. The code for the function RandomWalk[] is in the file **RandomWalk.m** in Appendix B.

■ A.1 Programming Exercises

Most of the exercises require you to modify the code given in the book or to expand its functionality. The references in parentheses point to the place in the book where the topic of the exercise is covered. The difficulty is rated with a number from 0 to 10 in square brackets with 0 being trivial and 10 being impossible.

■ **1.** [4] Modify `CartesianMap[]` and `PolarMap[]` in the package **ComplexMap.m** so that the two sets of lines (horizontal and vertical or radial and angular) are displayed in different colors or gray levels (Chapter 1).

■ **2.** [6] Write a definition for expanding powers using the binomial formula $(a+b)^n = \sum_{i=0}^{n} \binom{n}{i} a^i b^{n-i}$ that does not use any auxiliary variables (page 181).

■ **3.** [7] For the purpose of phototypesetting the manuscript for this book, the default width of lines in all kinds of graphics had to be changed. Find a way to do this for `Plot[]`, `Plot3D[]`, `ListPlot[]`, `Graphics[]`, and `Graphics3D[]`. This should be done so that all graphics will have thinner lines without giving some special command when producing them (Chapters 3 and 6).

■ **4.** [9] Find the fastest way of computing the nth Fibonacci number (page 71).

■ **5.** [6] Write a package implementing the properties of the Dirac delta function used in mathematical physics (Chapter 8).

■ **6.** [4] Add rules to **Abs.m** for definite integrals of `Abs[]` and `Sign[]` (page 187).

■ **7.** [4] If you define a rule like `f'[x_] := 1 + x^2` then `f''[x]` is not simplified to 2x as one might expect. Find a rule that will simplify higher derivatives (page 181).

■ **8.** [8] There is a difference between the built-in `ReadList[]` and the function `MyReadList[]` from subsection 9.2.6. If opening the file fails, `ReadList[]` returns itself unevaluated and `MyReadList[]` returns the empty list `{}`. Write a version of `MyReadList[]` that behaves like `ReadList[]` in this respect (page 202).

■ **9.** [3] The function `Explode[]` in subsection 6.5.4 is defined as

```
Explode[atom_] := Characters[ ToString[InputForm[atom]] ].
```

Can you explain why we used `ToString[InputForm[atom]]` instead of simply `ToString[atom]` (page 197)?

■ **10.** [7] Write a definition for the format type `TeXForm` for tensors without looking at the listing of **Tensors.m** in Appendix B (page 158).

■ A.2 Solutions to Exercises

We give short program fragments and hints only. It should be easy to turn those into a complete package if this is desired.

■ **1.** In the auxiliary procedure MakeLines[] maintain the horizontal and vertical lines separately and insert an appropriate graphics primitive (RGBColor[] or GrayLevel[], for example) at the beginning of each of these two sets of lines and then combine them.

■ **2.** Write the expression inside the Sum[] as a pure function and apply it to the range of integers from 0 to n and then add them together.

```
(a_ + b_)^n_Integer?Positive :=
    Plus @@ (Binomial[n, #] a^# b^(n-#)& /@ Range[0, n])
```

Functional form of expansion of powers

■ **3.** For Plot[] and ListPlot[], you can set the default option PlotStyle to Thickness[*val*]. For Plot3D[] the option is MeshStyle.

For Graphics[] and Graphics3D[], the following rule will prepend the graphics primitive Thickness[*val*] to all graphics that do not have it already.

```
Graphics[l_List] :=
    Graphics[Prepend[l, Thickness[0.001]]] /;
        Length[l] == 0 || Head[l[[1]]] =!= Thickness
```

■ **4.** There are formulae giving the nth Fibonacci number directly that can be evaluated to a numerical approximation sufficient for rounding to the correct integer. The fastest method known to the author uses the fact that the power $n-1$ of the matrix

$$\begin{pmatrix} 1 & 1 \\ 1 & 0 \end{pmatrix}$$

has the nth Fibonacci number in the top left corner. A power tree method for computing matrix powers is very fast.

```
fib[n_] :=
    Block[{ result = {{1,0},{0,1}}, new = n-1, m = {{1,1},{1,0}} },
        While[ new > 1,
            If[ OddQ[new], result = result . m ];
            new = Floor[new/2];
            m = m . m
        ];
    m[[1,1]] result[[1,1]] + m[[2,1]] result[[1,2]]
    ]
```

Fast computation of matrix powers

On a MIPS M/120 computer `fib[10^5]` takes just 36 seconds. The result is approximately $2.597406934722172 \cdot 10^{20898}$.

■ **5.** The most important properties are the integral formulae

$$\int_{-\infty}^{\infty} \delta(x)\mathrm{d}x = 1$$

$$\int_{-\infty}^{\infty} f(x)\delta(x-a)\mathrm{d}x = f(a)$$

which can be given as

```
DiracDelta/:
    Integrate[DiracDelta[x_], {x_, -Infinity, Infinity}] = 1
DiracDelta/:
    Integrate[f_[x_] DiracDelta[x_ + a_.], {x_, -Infinity, Infinity}] = f[-a]
```

The rule $\delta(x) = 0, x \neq 0$ is also straightforward:

```
DiracDelta[x_] := 0 /; x != 0.
```

This rule does not apply to symbolic arguments, which is important here!

■ **6.** The formula for expressing the definite integral in terms of the indefinite one is $\int_a^b f(x)\mathrm{d}x = F(b) - F(a)$, where $F(x)$ is the indefinite integral of $f(x)$.

```
Abs /: Integrate[Abs[x_], {x_, a_, b_}] := (b Abs[b] - a Abs[a])/2
Sign/: Integrate[Sign[x_], {x_, a_, b_}] := Abs[b] - Abs[a]
```

■ **7.** The reason the rule does not match is that the left side is `Derivative[1][f][x_]` and the expression `Derivative[2][f][x]` clearly does not match this. A possible rule is

$$\texttt{Derivative[n\_][f][x\_] := D[1 + x\textasciicircum 2, \{x, n-1\}]}.$$

■ **8.** You cannot simply say

$$\texttt{Return⌈MyReadList[fileName, thing]]}$$

inside `MyReadList[]` as this would lead to an infinite loop (the rule still matches). Instead, you can open the file as part of a side condition to the rule and simply let that condition fail if the open did not succeed. The rule will then not match and since there are no other rules for `MyReadList[]`, it will be left alone. The side condition can be put inside the block so that it can share local variables with the body of the rule. This is admittedly rather obscure.

```
MyReadList[fileName_String, thing_:Expression] :=
    Block[{file, expr, list = {}}, (
            While[ True,
                expr = Read[file, thing];
                If[ expr === EndOfFile, Break[] ];
                AppendTo[list, expr]
            ];
            Close[file];
            list
        ) /; (
            file = OpenRead[fileName];
            file =!= Null
        )
    ]
```

■ **9.** For symbols and integers it would indeed not make any difference. For other expressions `ToString[`*expr* `]` would give the string corresponding to the usual *two-dimensional* output form. It would be impossible to read this back in with `Intern[]`. If *expr* is a string itself, `ToString[`*expr* `]` would not include the quotation marks in the output string and `Intern[]` would convert the result back to a symbol rather than a string.

■ **10.** Since T_EX does not like multiple subscript or superscripts we have to insert something in between them. Usually one uses {} which produces no text.

```
Format[ Tensor[t_][ind___], TeXForm ] :=
    Block[{indices},
        indices = {ind} /. {ui->Superscript, li->Subscript};
        indices = Transpose[{Table["{}", {Length[indices]}], indices}];
        SequenceForm[t, Sequence @@ Flatten[indices, 1]]
    ]
```

We can still use the primitives `SubScript` and `Superscript` since *Mathematica* knows how to generate T_EX output for subscripts and superscripts. Our example

```
TeXForm[Tensor[Gamma][li[k], ui[i], ui[j]][x, y, z, t]]
```

now becomes `{\rm Gamma}{}_{k}{}^{i}{}^{j}(x,y,z,t)`. When run through T_EX it produces $\text{Gamma}_k{}^{ij}(x, y, z, t)$ and we notice that we should have defined a T_EX form for `Gamma` with

```
Format[Gamma, TeXForm] = "\\Gamma",
```

to get `\Gamma{}_{k}{}^{i}{}^{j}(x,y,z,t)` or finally $\Gamma_k{}^{ij}(x, y, z, t)$.

Program Listings

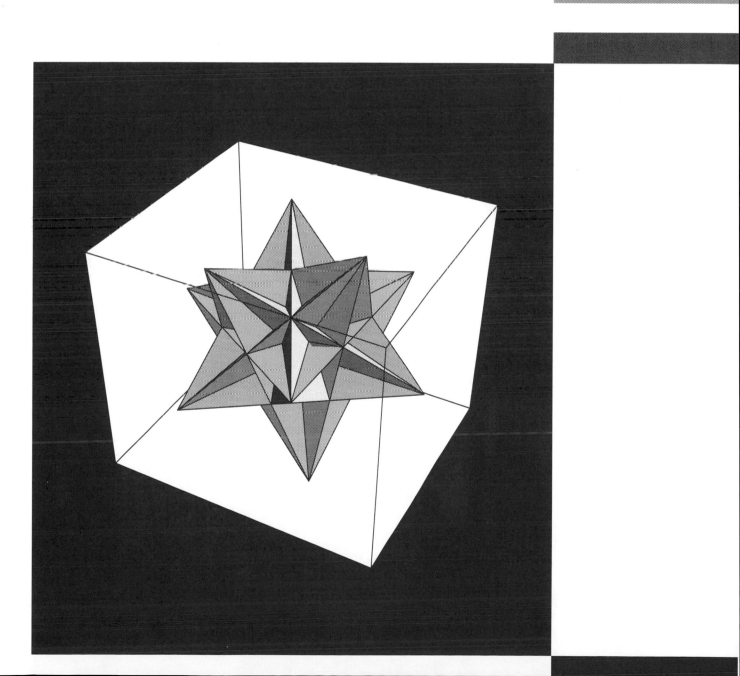

This appendix contains an index to all the packages developed in this book. Packages that do not appear in full in one of the chapters are printed in this appendix as well. All these packages are available in machine-readable form from many sources.

The programs listed here do not contain as many comments as they should. My excuse is that everything is explained in great detail in this book. Of course, you should put many more comments into your own programs.

About the illustration overleaf:

The *great icosahedron* is another one of the regular polyhedra with self-intersections. We saw the others in section 4.5. The great icosahedron—like the icosahedron itself—consists of 20 regular triangles. Its vertices can be derived from those of the icosahedron.

■ B.1 Index to Complete Listings of All Packages

This table lists the pages on which the final versions of the corresponding packages appear. The ones that were not reproduced in full in one of the chapters appear on the following pages.

All these packages are available in machine-readable form. For the Macintosh and for MS-DOS systems, they can be ordered from Wolfram Research, Inc. For other computers, they are available through the *Mathematica* user group and through many electronic bulletin boards.

Package	Page
AlgExp.m	253
Atoms.m	158
BookPictures.m	255
Collatz.m	206
ComplexMap.m	242
ComplexTest.m	27
Constants.m	166
CoverPage.m	229
CylindricalPlot3D.m	41
ExpandIt.m	52
FilterOptions.m	64
GetNumber.m	76
MakeFunctions.m	124
MyReadList.m	202
NewParametricPlot3D.m	246
Newton.m	85
Numerical.m	170
OptionUse.m	61
PlotUtilities.m	226
PrimePi.m	73
PrintTime.m	115
RandomWalk.m	254
ReIm.m	243
RungeKutta.m	252
SessionLog.m	208
ShowTime.m	183
Skeleton.m	48
Struve.m	251
SunView.m	126
SwinnertonDyer.m	105
Tensors.m	253
Transcript.m	211
TrigSimplification.m	249
Until.m	75
VectorCalculus.m	248
WrapHold.m	117

■ B.2   Additional Program Listings

■ B.2.1   ComplexMap.m

This package was developed in Chapter 1.

```
BeginPackage["ComplexMap`"]

CartesianMap::usage = "CartesianMap[f, {x0, x1, (dx)}, {y0, y1, (dy)}, options...] plots
    the image of the cartesian coordinate lines under the function f.
    The default values of dx and dy are chosen so that the number of lines
    is equal to the value of the option PlotPoints of Plot3D[]"

PolarMap::usage = "PolarMap[f, {r0:0, r1, (dr)}, {phi0, phi1, (dphi)}, options...] plots
    the image of the polar coordinate lines under the function f.
    The default values of dr and dphi are chosen so that the number of lines
    is equal to the value of the option PlotPoints of Plot3D[].
    The default for the phi range is {0, 2Pi}."

Begin["`Private`"]

huge = 10.0^6

SplitLine[vl_] :=
    Block[{vll, pos, linelist = {}, low, high},
        vll = If[NumberQ[#], #, Indeterminate]& /@ vl;
        pos = Flatten[ Position[vll, Indeterminate] ];
        pos = Union[ pos, {0, Length[vll]+1} ];
        Do[ low = pos[[i]]+1;
            high = pos[[i+1]]-1;
            If[ low < high, AppendTo[linelist, Take[vll, {low, high}]] ],
            {i, 1, Length[pos]-1}];
        linelist
    ]

MakeLines[points_] :=
    Block[{lines, newpoints},
        newpoints = points /.
            { z_?NumberQ :> huge z/Abs[z] /; Abs[z] > huge,
              z_?NumberQ :> 0.0 /; Abs[z] < 1/huge,
              DirectedInfinity[z_] :> huge z/Abs[z] };
        lines = Join[ newpoints, Transpose[newpoints] ];
        lines = Flatten[ SplitLine /@ lines, 1 ];
        lines = Map[ {Re[#], Im[#]}&, lines, {2} ];
        lines = Map[ Line, lines ];
        Graphics[lines]
    ]

FilterOptions[ command_Symbol, opts___ ] :=
    Block[{keywords = First /@ Options[command]},
        Sequence @@ Select[ {opts}, MemberQ[keywords, First[#]]& ]
```

```
        ]

CartesianMap[ f_, {x0_, x1_, dx_:Automatic}, {y0_, y1_, dy_:Automatic}, opts___ ] :=
    Block[ {x, y, points, plotpoints, ndx=N[dx], ndy=N[dy]},
        plotpoints = PlotPoints /. {opts} /. Options[Plot3D];
        If[ dx === Automatic, ndx = N[(x1-x0)/(plotpoints-1)] ];
        If[ dy === Automatic, ndy = N[(y1-y0)/(plotpoints-1)] ];
        points = Table[ N[f[x + I y]],
            {x, x0, x1, ndx}, {y, y0, y1, ndy} ];
        Show[ MakeLines[points], FilterOptions[Graphics, opts],
            AspectRatio->Automatic, Axes->Automatic ]
    ] /; NumberQ[N[x0]] && NumberQ[N[x1]] && NumberQ[N[y0]] && NumberQ[N[y1]]
        (NumberQ[N[dx]] || dx === Automatic) &&
        (NumberQ[N[dy]] || dy === Automatic)

PolarMap[ f_, {r0_:0, r1_, dr_:Automatic}, {phi0_, phi1_, dphi_:Automatic}, opts___ ] :=
    Block[ {r, phi, points, plotpoints, ndr=dr, ndphi=dphi},
        plotpoints = PlotPoints /. {opts} /. Options[Plot3D];
        If[ dr === Automatic, ndr = N[(r1-r0)/(plotpoints-1)] ];
        If[ dphi === Automatic, ndphi = N[(phi1-phi0)/(plotpoints-1)] ];
        points = Table[ N[f[r Exp[I phi]]],
            {r, r0, r1, ndr}, {phi, phi0, phi1, ndphi} ];
        Show[ MakeLines[points], FilterOptions[Graphics, opts],
            AspectRatio->Automatic, Axes->Automatic ]
    ] /; NumberQ[N[r0]] && NumberQ[N[r1]] && NumberQ[N[phi0]] && NumberQ[N[phi1]]
        (NumberQ[N[dr]] || dr === Automatic) &&
        (NumberQ[N[dphi]] || dphi === Automatic)

PolarMap[ f_, rRange_List, opts___Rule ] :=
    PolarMap[ f, rRange, {0, 2Pi}, opts ]

End[ ]

Protect[CartesianMap, PolarMap]

EndPackage[ ]
```

■ B.2.2 Relm.m

This is an improved version of the standard package **Algebra/Relm.m**.

```
(*  to declare a variable x to be real, define    x/: Im[x] = 0  *)

BeginPackage["ReIm`"]

(* no exports *)

Begin["`Private`"]

protected = Unprotect[Re,Im, Abs, Conjugate, Arg]
```

```
(* fundamental rules, Im[x]==0 serves as our test for 'reality' *)

Re[x_] := x  /; Im[x] == 0
Conjugate[x_] :=  x /; Im[x] == 0
Conjugate[x_] := -x /; Re[x] == 0

(* there must not be a rule for Im[x] in terms of Re[x] !! *)

(* things known to be real *)

Im[Re[ _ ]]  := 0
Im[Im[ _ ]]  := 0
Im[Abs[ _ ]] := 0
Im[Arg[ _ ]] := 0

(* arithmetic *)

Re[x_ + y_] := Re[x] + Re[y]
Im[x_ + y_] := Im[x] + Im[y]

Re[x_ y_] := Re[x] Re[y] - Im[x] Im[y]
Im[x_ y_] := Re[x] Im[y] + Im[x] Re[y]

Re[ 1/x_ ] :=  Re[x] / (Re[x]^2 + Im[x]^2)
Im[ 1/x_ ] := -Im[x] / (Re[x]^2 + Im[x]^2)

Re[E^x_] := Cos[Im[x]] Exp[Re[x]]
Im[E^x_] := Sin[Im[x]] Exp[Re[x]]

Im[x_^2] := 2 Re[x] Im[x]

Re[ x_^n_Integer ] :=
    Block[{a, b},
        a = Round[n/2]; b = n-a;
        Re[x^a] Re[x^b] - Im[x^a] Im[x^b]
    ]

Im[ x_^n_Integer ] :=
    Block[{a, b},
        a = Round[n/2]; b = n-a;
        Re[x^a] Im[x^b] + Im[x^a] Re[x^b]
    ]

Re[x_Integer^n_Rational] := 0                    /; IntegerQ[2n] && x<0
Im[x_Integer^n_Rational] :=
    (-x)^n (-1)^((Numerator[n]-1)/2) /; IntegerQ[2n] && x<0

Re[x_Integer^n_Rational] := x^n /; OddQ[Denominator[n]] || x>0
Im[x_Integer^n_Rational] := 0   /; OddQ[Denominator[n]] || x>0

Re[(-1)^n_Rational] := Cos[n Pi]
Im[(-1)^n_Rational] := Sin[n Pi]

(* functions *)

Re[Sin[x_]] := Sin[Re[x]] Cosh[Im[x]]
Im[Sin[x_]] := Cos[Re[x]] Sinh[Im[x]]
```

```
Re[Cos[x_]] :=  Cos[Re[x]] Cosh[Im[x]]
Im[Cos[x_]] := -Sin[Re[x]] Sinh[Im[x]]

Re[Log[r_?Positive]] := Log[r]
Im[Log[r_?Positive]] := 0
Re[Log[r_?Negative]] := Log[-r]
Im[Log[r_?Negative]] := Pi

Re[Log[z_]] := (1/2) Log[Re[z]^2 + Im[z]^2]
Im[Log[z_]] := Arg[z]

Re[Log[a_ b_]] := Re[Log[a] + Log[b]]
Im[Log[a_ b_]] := Im[Log[a] + Log[b]]
Re[Log[a_^c_]] := Re[c Log[a]]
Im[Log[a_^c_]] := Im[c Log[a]]

Re[Tan[x_]] := Re[Sin[x]/Cos[x]]
Im[Tan[x_]] := Im[Sin[x]/Cos[x]]

(* conjugates *)

Re[Conjugate[z_]] :=  Re[z]
Im[Conjugate[z_]] := -Im[z]

Conjugate[x_+y_]:= Conjugate[x]+Conjugate[y]
Conjugate[x_ y_]:= Conjugate[x] Conjugate[y]
Conjugate[x_^n_Integer]:= Conjugate[x]^n
Conjugate[Conjugate[x_]]:= x

Protect[Release[protected]]

End[]
EndPackage[]
```

■ B.2.3 NewParametricPlot3D.m

NewParametricPlot3D.m is an extension of the standard package **ParametricPlot3D.m** that we developed in subsection 11.1.1. It includes SphericalPlot3D[] from subsection 2.2.4 and has some additional features needed for the picture on the book cover. See Chapter 11.

```
BeginPackage["ParametricPlot3D`"]

ParametricPlot3D::usage =
    "ParametricPlot3D[{x,y,z,(style)}, {u,u0,u1,(du)}, {v,v0,v1,(dv)}, (options..)]
    plots a 3D parametric surface. Options are passed to Show[]"

PointParametricPlot3D::usage =
    "PointParametricPlot3D[{x,y,z}, {u,u0,u1,(du)}, {v,v0,v1,(dv)}, (options..)]
    plots a two-parameter set of points in space. Options are passed to Show[]."

SpaceCurve::usage = "SpaceCurve[{x,y,z}, {u,u0,u1,(du)}, (options..)]
    plots a 3D parametric curve. Options are passed to Show[]."

PointSpaceCurve::usage = "PointSpaceCurve[{x,y,z}, {u,u0,u1,(du)}, (options..)]
    plots a one-parameter set of points in space. Options are passed to Show[]"

SphericalPlot3D::usage = "SphericalPlot3D[r, {theta-range}, {phi-range}, (options...)]
    plots r as a function of the angles theta and phi.
    SphericalPlot3D[{r, style}, ...] uses style to render each surface patch"

CylindricalPlot3D::usage = "CylindricalPlot3D[z, {r-range}, {phi-range}, (options...)]
    plots z as a function of r and phi.
    CylindricalPlot3D[{z, style},  ...] uses style to render each surface patch"

Begin["`Private`"]

MakePolygons[vl_List] :=
    Block[{l = vl, l1 = Map[RotateLeft, vl], mesh},
        mesh = {l, RotateLeft[l], RotateLeft[l1], l1};
        mesh = Map[Drop[#, -1]&, mesh, {1}];
        mesh = Map[Drop[#, -1]&, mesh, {2}];
        Polygon /@ Transpose[ Map[Flatten[#, 1]&, mesh] ]
    ]  /; TensorRank[vl] == 3 && Dimensions[vl][[3]] == 3

MakePolygons[vl_List] :=
    Block[{l, l1, mesh, cols},
        l = Map[Take[#, 3]&, vl, {2}]; (* the coords *)
        l1 = Map[RotateLeft, l];
        cols = Map[#[[4]]&, vl, {2}];  (* the colors *)
        mesh = {l, RotateLeft[l], RotateLeft[l1], l1};
        mesh = Map[Drop[#, -1]&, mesh, {1}];
        mesh = Map[Drop[#, -1]&, mesh, {2}];
        cols = Drop[cols, -1];
        cols = Map[Drop[#, -1]&, cols, {1}];
        mesh = Polygon /@ Transpose[ Map[Flatten[#, 1]&, mesh] ];
```

```
                Flatten[Transpose[{Flatten[cols], mesh}]]
        ]  /; TensorRank[vl] == 3 && Dimensions[vl][[3]] == 4

FilterOptions[ command_Symbol, opts___ ] :=
    Block[{keywords = First /@ Options[command]},
        Sequence @@ Select[ {opts}, MemberQ[keywords, First[#]]& ]
    ]

Attributes[ParametricPlot3D] = {HoldFirst}

ParametricPlot3D[ fun_,
        {u_, u0_, u1_, du_:Automatic}, {v_, v0_, v1_, dv_:Automatic}, opts___ ] :=
    Block[{plotpoints, ndu = N[du], ndv = N[dv]},
        plotpoints = PlotPoints /. {opts} /. Options[Plot3D];
        If[ du === Automatic, ndu = N[(u1-u0)/(plotpoints-1)] ];
        If[ dv === Automatic, ndv = N[(v1-v0)/(plotpoints-1)] ];
        Show[ Graphics3D[MakePolygons[Table[ N[fun],
                                            {u, u0, u1, ndu}, {v, v0, v1, ndv}] ]],
            FilterOptions[Graphics3D, opts] ]
    ]  /; NumberQ[N[u0]] && NumberQ[N[u1]] && NumberQ[N[v0]] && NumberQ[N[v1]]

Attributes[PointParametricPlot3D] = {HoldFirst}

PointParametricPlot3D[ fun_,
        {u_, u0_, u1_, du_:Automatic}, {v_, v0_, v1_, dv_:Automatic}, opts___ ] :=
    Block[{plotpoints, ndu = N[du], ndv = N[dv]},
        plotpoints = PlotPoints /. {opts} /. Options[Plot3D];
        If[ du === Automatic, ndu = N[(u1-u0)/(plotpoints-1)] ];
        If[ dv === Automatic, ndv = N[(v1-v0)/(plotpoints-1)] ];
        Show[ Graphics3D[Table[ Point[N[fun]], {u, u0, u1, ndu}, {v, v0, v1, ndv} ]],
            FilterOptions[Graphics3D, opts] ]
    ]  /; NumberQ[N[u0]] && NumberQ[N[u1]] && NumberQ[N[v0]] && NumberQ[N[v1]]

Attributes[SpaceCurve] = {HoldFirst}

SpaceCurve[ fun_, ul:{_, u0_, u1_, du_}, opts___ ] :=
    Show[ Graphics3D[Line[Table[ N[fun], ul ]]], FilterOptions[Graphics3D, opts] ] /;
        NumberQ[N[u0]] && NumberQ[N[u1]] && NumberQ[N[du]]

SpaceCurve[ fun_, {u_, u0_, u1_}, opts___ ] :=
    Block[{plotpoints},
        plotpoints = PlotPoints /. {opts} /. Options[Plot3D];
    SpaceCurve[ fun, {u, u0, u1, (u1-u0)/(plotpoints-1)}, opts ]
    ]

Attributes[PointSpaceCurve] = {HoldFirst}

PointSpaceCurve[ fun_, ul:{_, u0_, u1_, du_}, opts___ ] :=
    Show[ Graphics3D[Table[ Point[N[fun]], ul ]], FilterOptions[Graphics3D, opts] ] /;
        NumberQ[N[u0]] && NumberQ[N[u1]] && NumberQ[N[du]]

PointSpaceCurve[ fun_, {u_, u0_, u1_}, opts___ ] :=
```

```
    Block[{plotpoints},
        plotpoints = PlotPoints /. {opts} /. Options[Plot3D];
        PointSpaceCurve[ fun, {u, u0, u1, (u1-u0)/(plotpoints-1)}, opts ]
    ]

Attributes[SphericalPlot3D] = {HoldFirst}

SphericalPlot3D[ {r_, style_}, tlist:{theta_, __}, plist:{phi_, __}, opts___ ] :=
    Block[{rs},
        ParametricPlot3D[ {(rs = r) Sin[theta] Cos[phi],
                           rs Sin[theta] Sin[phi],
                           rs Cos[theta],
                           style},
                           tlist, plist, opts ]
    ]

SphericalPlot3D[ r_, tlist:{theta_, __}, plist:{phi_, __}, opts___ ] :=
    ParametricPlot3D[ r{Sin[theta] Cos[phi],
                        Sin[theta] Sin[phi],
                        Cos[theta]},
                      tlist, plist, opts ]


Attributes[CylindricalPlot3D] = {HoldFirst}

CylindricalPlot3D[ {z_, style_}, rlist:{r_, __}, plist:{phi_, __}, opts___ ] :=
    ParametricPlot3D[{r Cos[phi], r Sin[phi], z, style}, rlist, plist, opts]

CylindricalPlot3D[ z_, rlist:{r_, __}, plist:{phi_, __}, opts___ ] :=
    ParametricPlot3D[{r Cos[phi], r Sin[phi], z}, rlist, plist, opts]

End[]

Protect[ParametricPlot3D, PointParametricPlot3D, SpaceCurve,
    PointSpaceCurve, SphericalPlot3D, CylindricalPlot3D]

EndPackage[]
```

■ B.2.4 VectorCalculus.m

Using inner and outer products for vector calculus was discussed in subsection 4.6.3.

```
BeginPackage["VectorCalculus`"]

Div::usage = "Div[v, varlist] computes the divergence of
    the vectorfield v w.r.t. the given variables in Cartesian coordinates."

Laplacian::usage = "Laplacian[s, varlist] computes the Laplacian of
    the scalar field s w.r.t. the given variables in Cartesian coordinates."

Grad::usage = "Grad[s, varlist] computes the gradient of s
```

w.r.t. the given variables in Cartesian coordinates."

JacobianMatrix::usage = "JacobianMatrix[flist, varlist] computes the Jacobian of
    the functions flist w.r.t. the given variables."

Begin["`Private`"]

Div[v_List, var_List] := Inner[D, v, var, Plus]

Grad[s_, var_List] := D[s, #]& /@ var

Laplacian[s_, var_List] := Div[Grad[s, var], var]

JacobianMatrix[f_List, var_List] := Outer[D, f, var]

End[]

EndPackage[]

■ B.2.5 TrigSimplification.m

This package was derived in subsection 6.2. It provides basically the same functionality
as the standard package **Trigonometry.m**.

```
BeginPackage["TrigSimplification`"]

TrigNormal::usage = "TrigNormal[e] puts expressions with trigonometric
    functions into normal form."

TrigLinear::usage = "TrigLinear[e] expands products and powers of trigonometric
    functions."

TrigArgument::usage = "TrigArgument[e] writes trigonometric functions of multiple
    angles as products of functions of that angle."

Begin["`Private`"]

TrigCanonicalRules = {
    Sin[n_?Negative x_.] :> -Sin[-n x],
    Cos[n_?Negative x_.] :>  Cos[-n x],
    Tan[n_?Negative x_.] :> -Tan[-n x],
    Sin[n_?Negative x_ + y_] :> - Sin[-n x - y] /; OrderedQ[{x, y}],
    Cos[n_?Negative x_ + y_] :>   Cos[-n x - y] /; OrderedQ[{x, y}],
    Tan[n_?Negative x_ + y_] :> - Tan[-n x - y] /; OrderedQ[{x, y}]
}

TrigLinearRules = {
    Sin[x_] Sin[y_] :> Cos[x-y]/2 - Cos[x+y]/2,
    Cos[x_] Cos[y_] :> Cos[x+y]/2 + Cos[x-y]/2,
    Sin[x_] Cos[y_] :> Sin[x+y]/2 + Sin[x-y]/2,
    Sin[x_]^(m1_Integer?EvenQ) :> Block[{m=Abs[m1]},
      (2^(-m+1)
        (Sum[(-1)^(m/2-k) Binomial[m,k] Cos[(m-2k)x], {k, 0, m/2-1}]+
```

```
        Binomial[m,m/2]/2))^Sign[m1] ],
    Cos[x_]^(m1_Integer?EvenQ) :> Block[{m=Abs[m1]},
     (2^(-m+1)
      (Sum[Binomial[m,k] Cos[(m-2k)x], {k, 0, m/2-1}] +
       Binomial[m,m/2]/2))^Sign[m1] ],
    Sin[x_]^(m1_Integer?OddQ) :> Block[{m=Abs[m1]},
     (2^(-m+1)
      Sum[(-1)^((m-1)/2-k)
         Binomial[m,k] Sin[(m-2k)x], {k, 0, (m-1)/2}])^Sign[m1] ],
    Cos[x_]^(m1_Integer?OddQ) :> Block[{m=Abs[m1]},
     (2^(-m+1)
      Sum[Binomial[m,k] Cos[(m-2k)x], {k, 0, (m-1)/2}])^Sign[m1] ]
}
TrigArgumentRules = {
    Sin[x_ + y_] :> Sin[x] Cos[y] + Sin[y] Cos[x],
    Cos[x_ + y_] :> Cos[x] Cos[y] - Sin[x] Sin[y],
    Sin[n_Integer?Positive x_.] :>
        Sum[ (-1)^((i-1)/2) Binomial[n, i] Cos[x]^(n-i) Sin[x]^i,
            {i, 1, n, 2} ],
    Cos[n_Integer?Positive x_.] :>
        Sum[ (-1)^(i/2) Binomial[n, i] Cos[x]^(n-i) Sin[x]^i,
            {i, 0, n, 2} ]
}

SetAttributes[{TrigNormal, TrigLinear, TrigArgument}, Listable]

TrigNormal[e_] := e /. TrigCanonicalRules

TrigLinear[e_] :=
    FixedPoint[ Expand[# //. TrigLinearRules /. TrigCanonicalRules]&, e ]

TrigArgument[e_] :=
    Together[ FixedPoint[ (# //. TrigArgumentRules /. TrigCanonicalRules)&, e ] ]

End[]

Protect[TrigNormal, TrigLinear, TrigArgument]

EndPackage[]
```

■ B.2.6 Struve.m

The topic of subsection 8.4, this package shows how to implement a new mathematical function in *Mathematica*.

```
BeginPackage["Struve`"]

StruveH::usage "StruveH[nu, z] gives the Struve function."

Begin["`Private`"]

Attributes[StruveH] = {Listable}

(* special values *)

StruveH[r_Rational?Positive, z_] :=
    BesselY[r, z] +
    Sum[Gamma[m + 1/2] (z/2)^(-2m + r - 1)/Gamma[r + 1/2 - m], {m, 0, r-1/2}]/Pi /;
        Denominator[r] == 2

StruveH[r_Rational?Negative, z_] :=
    (-1)^(-r-1/2) BesselJ[-r, z] /; Denominator[r] == 2

(* Series expansion *)

StruveH/: Series[StruveH[nu_?NumberQ, z_], {z_, 0, ord_Integer}] :=
    (z/2)^(nu + 1) Sum[ (-1)^m (z/2)^(2m)/Gamma[m + 3/2]/Gamma[m + nu + 3/2],
                    {m, 0, (ord-nu-1)/2} ] + O[z]^(ord+1)

(* numerical evaluation *)

StruveH[nu_?NumberQ, z_?NumberQ] :=
    Block[{s=0, so=-1, m=0, prec = Precision[z],
            z2 = -(z/2)^2,k1 = 3/2, k2 = nu + 3/2, g1, g2, zf},
        zf = (z/2)^(nu+1); g1 = Gamma[k1]; g2 = Gamma[k2];
        While[so != s,
            so = s; s += N[zf/g1/g2, prec];
            g1 *= k1; g2 *= k2; zf *= z2;
            k1++; k2++; m++
        ];
        s
    ]

(* derivatives *)

Derivative[0, n_Integer?Positive][StruveH][nu_, z_] :=
    D[ (StruveH[nu-1, z] - StruveH[nu+1, z] + (z/2)^nu/Sqrt[Pi]/Gamma[nu + 3/2])/2,
        {z, n-1} ]

End[]

Protect[StruveH]

EndPackage[]
```

■ B.2.7 RungeKutta.m

This is the final version of the packages **RK1.m** and **RK2.m** from subsection 7.3.

```
BeginPackage["RungeKutta`"]

RungeKutta::usage = "RungeKutta[{e1,e2,..}, {y1,y2,..}, {a1,a2,..}, {t1, dt}]
    numerically integrates the ei as functions of the yi with inital values ai.
    The integration proceeds in steps of dt from 0 to t1
    RungeKutta[{e1,e2,..}, {y1,y2,..}, {a1,a2,..}, {t, t0, t1, dt}] integrates
    a time-dependent system from t0 to t1."

Begin["`Private`"]

RKStep[f_, y_, y0_, dt_] :=
    Block[{ k1, k2, k3, k4 },
        k1 = dt N[ f /. Thread[y -> y0] ];
        k2 = dt N[ f /. Thread[y -> y0 + k1/2] ];
        k3 = dt N[ f /. Thread[y -> y0 + k2/2] ];
        k4 = dt N[ f /. Thread[y -> y0 + k3] ];
        y0 + (k1 + 2 k2 + 2 k3 + k4)/6
    ]

RungeKutta[f_List, y_List, y0_List, {t1_, dt_}] :=
    NestList[ RKStep[f, y, #, N[dt]]&, N[y0], Round[N[t1/dt]] ] /;
        Length[f] == Length[y] == Length[y0]

RungeKutta[f_List, y_List, y0_List, {t_, t0_, t1_, dt_}] :=
    Block[{res},
        res = RungeKutta[ Append[f, 1], Append[y, t], Append[y0, t0], {t1 - t0, dt} ];
        Drop[#, -1]& /@ res
    ] /; Length[f] == Length[y] == Length[y0]

End[]

Protect[RungeKutta]

EndPackage[]
```

■ B.2.8 AlgExp.m

The predicate `AlgExpQ[]` was developed in subsection 6.5. Here it is with added messages and a minimal package context.

```
AlgExpQ::usage = "AlgExpQ[expr] returns true if expr is an algebraic expression."

Begin["Private`"]

SetAttributes[AlgExpQ, Listable]

AlgExpQ[ _Integer ]  = True
AlgExpQ[ _Rational ] = True
AlgExpQ[ c_Complex ] = AlgExpQ[Re[c]] && AlgExpQ[Im[c]]
AlgExpQ[ _Symbol ]   = True

AlgExpQ[ a_ + b_ ] := AlgExpQ[a] && AlgExpQ[b]
AlgExpQ[ a_ * b_ ] := AlgExpQ[a] && AlgExpQ[b]
AlgExpQ[ a_ ^ b_Integer ]  := AlgExpQ[a]
AlgExpQ[ a_ ^ b_Rational ] := AlgExpQ[a]

AlgExpQ[_] = False

End[]
Null
```

■ B.2.9 Tensors.m

Tensors.m contains some definitions for formatting tensors with upper and lower indices. It was developed in subsection 9.1.3.

```
BeginPackage["Tensors`"]

ui::usage = "ui[index] denotes an upper index in a tensor."
li::usage = "li[index] denotes a lower index in a tensor."
Tensor::usage = "Tensor[h][indices] denotes a tensor h with index list indices."

Begin["`Private`"]

Format[ Tensor[t_][ind___] ] :=
    Block[{indices},
        indices = {ind} /. {ui->Superscript, li->Subscript};
        SequenceForm[t, Sequence @@ indices]
    ]

Format[ Tensor[t_][ind___], TeXForm ] :=
    Block[{indices},
        indices = {ind} /. {ui->Superscript, li->Subscript};
        indices = Transpose[{Table["{}", {Length[indices]}], indices}];
        SequenceForm[t, Sequence @@ Flatten[indices, 1]]
    ]
```

```
End[]

EndPackage[]
```

■ B.2.10 RandomWalk.m

This little function was used for the picture at the beginning of Appendix A.

```
BeginPackage["RandomWalk`"]

RandomWalk::usage = "RandomWalk[n] plots a random walk of length n."

Begin["`Private`"]

RandomWalk[n_Integer] :=
    Block[{loc = {0.0, 0.0}, dir, points = Table[0, {n+1}], range = N[{0, 2Pi}]},
        points[[1]] = loc;
        Do[
            dir = Random[Real, range];
            loc += {Cos[dir], Sin[dir]};
            points[[i]] = loc,
          {i, 2, n+1}];
        Show[ Graphics[{Point[{0,0}], Line[points]}],
            Framed->True, AspectRatio->Automatic ]
    ]

End[]

EndPackage[]
```

■ B.2.11 BookPictures.m

This file contains the code for most of the pictures used for the chapter titles. See
subsection 5.6.3.

```
Needs["ComplexMap`"]
Needs["ParametricPlot3D`", "NewParametricPlot3D.m"]
Needs["Graphics`Shapes`"]
Needs["Graphics`Polyhedra`"]
Needs["RungeKutta`"]
Needs["RandomWalk`"]

(* Moebius transform *)

chapter1 := PolarMap[ (2#-I)/(#-1)&, {0.001, 5.001, 0.25}, {0, 2Pi, Pi/15},
                      Framed->True ]

(* Minimal surface *)

chapter2 :=
    ParametricPlot3D[{r*Cos[phi] - (r^2*Cos[2*phi])/2,
        -(r*Sin[phi]) - (r^2*Sin[2*phi])/2, (4*r^(3/2)*Cos[(3*phi)/2])/3},
        {r, 0.0001, 1, 0.9999/8}, {phi, 0, 4Pi, Pi/12}]

(* rotaionally symmetric parametric surface *)

chapter3 :=
    ParametricPlot3D[
    {r Cos[Cos[r]] Cos[psi], r Cos[Cos[r]] Sin[psi], r Sin[Cos[r]]},
    {r, 0.001, 9Pi/2 + 0.001, Pi/16}, {psi, 0, 3Pi/2, Pi/16}]

(* Fractal tile *)

om7 = N[-1+Sqrt[-3]]/2; l7=om7-2
r7 = {0, 1,-1,om7,-om7,om7+1,-om7-1}
g7[x_] := Flatten[Outer[Plus, r7 , l7 x]]

chapter4 := (
    points = Point[{Re[#],Im[#]}]& /@ Nest[g7, {0.}, 5];
    graph4 = Graphics[Prepend[points, PointSize[0.003]]];
    Show[graph4, AspectRatio->1, Axes->None, Framed->True]
    )

(* Sphere with random holes *)

chapter5 := Show[ Graphics3D[Select[Sphere[][[1]], Random[]>0.5&]] ]

(* Saddle surface *)

chapter6 := CylindricalPlot3D[r^2 Cos[2 phi],
        {r, 0, 1/2, 1/16}, {phi, 0, 2Pi, 2Pi/24}]

(* Van-der-Pol equation *)

chapter7 :=
```

```
      Block[{vdp, eps = 1.5, x, xdot},
          vdp = RungeKutta[{xdot, eps (1 - x^2) xdot - x}, {x, xdot},
                          {2, 0}, {4Pi, 0.05}];
          ListPlot[vdp, PlotJoined->True, AspectRatio -> Automatic]
      ]
```

```
(* Fourier approximations of saw-tooth *)
```

```
l5 = Table[ Sum[Sin[i x]/i, {i, n}], {n, 5} ];
```

```
chapter9 := Plot[ Release[l5], {x, -0.3, 2Pi+0.3}, Framed->True ]
```

```
(* spiral with varying radius *)
```

```
chapter10 :=
    ParametricPlot3D[{r (1 + phi/2) Cos[phi], r (1 + phi/2) Sin[phi], -phi/2},
        {r, 0.1, 1.1, 0.125}, {phi, 0, 11Pi/2, Pi/12}]
```

```
(* diagonally shaded surface *)
```

```
chapter11 :=
    SphericalPlot3D[ {Sin[theta],
        FaceForm[GrayLevel[0.05 + 0.9 Sin[2theta + phi]^2],
                GrayLevel[0.05 + 0.9 Sin[2theta - phi]^2]]},
        {theta, 0, Pi, Pi/24}, {phi, 0, 3Pi/2, Pi/12},
        Lighting->False ]
```

```
(* Random walk *)
```

```
appendixA := RandomWalk[5000]
```

```
(* Great icosahedron, vertices computed from icosahedron *)
```

```
AdjacentTo[face_, flist_] := Select[flist, Length[Intersection[face, #]] == 2&]
```

```
Opposite[face_, flist_] :=
    Block[ {adjacent, next},
        adjacent = AdjacentTo[ face, flist ];
        next = AdjacentTo[#, flist]& /@ adjacent;
        next = Complement[#, {face}]& /@ next;
        Flatten[ Intersection @@ #& /@ next ]
    ]
```

```
AppendTo[Polyhedra, GreatIcosahedron]
```

```
GreatIcosahedron/: Vertices[GreatIcosahedron] = Vertices[Icosahedron];
```

```
GreatIcosahedron/: Faces[GreatIcosahedron] =
    Opposite[#, Faces[Icosahedron]]& /@ Faces[Icosahedron];
```

```
appendixB := Show[ Polyhedron[GreatIcosahedron] ]
```

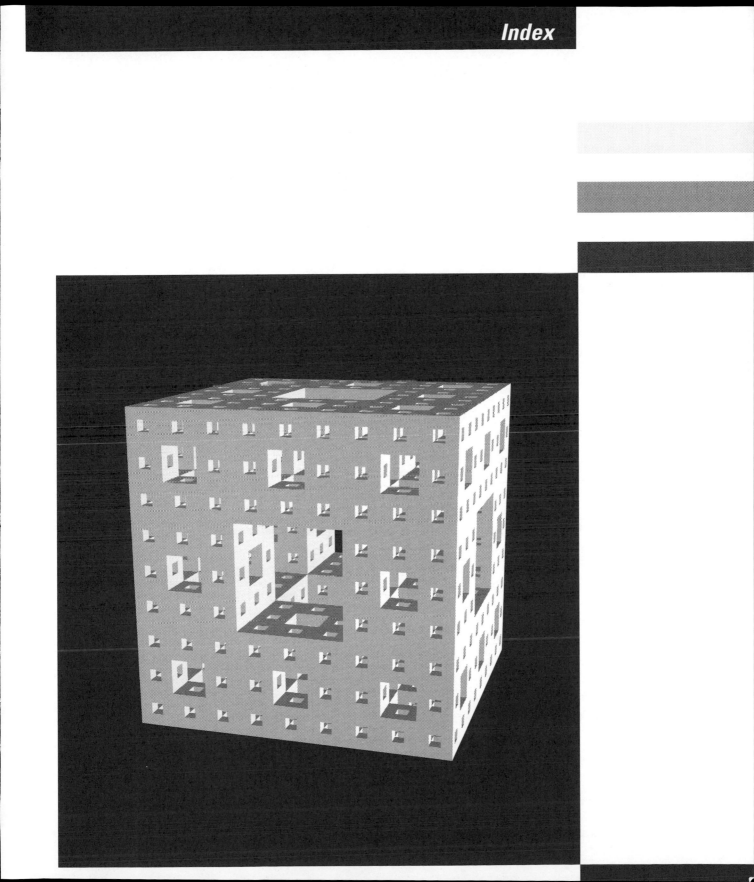

This index is not a substitute for the index in the *Mathematica* book. *Mathematica* commands are only listed here if they are treated in some detail. On the other hand you will find here all the commands developed in one of the example packages. These commands are not built-in and therefore not listed in the *Mathematica* book.

The main entry for a topic is set in boldface. If an entry stretches over several pages, only the first page is given. For typographical conventions, see page xiii.

About the illustration overleaf:

The third iteration of a Sierpinsky sponge, a 3-dimensional fractal figure. It is obtained by dividing a cube into 27 smaller cubelets and removing the six cubelets in the center of the faces and the one in the middle. Each of the remaining cubelets is then subdivided in the same way.

Software Information

Available Versions of *Mathematica* as of September 1, 1989

Macintosh Versions

- Standard Version: $495
- Enhanced Version: $795

4 megabytes recommended. (Less memory is required if a virtual memory system is used.)

386-Based MS-DOS Versions

- 386 Version: $695 (no floating-point coprocessor required)
- 386/7 Version: $995 (287 or 387 coprocessor required)
- 386/Weitek Version: $1,295 (Weitek coprocessor required)

640K memory and 1 megabyte extended memory required. Supports CGA, FGA, VGA, MCGA, 8514, and Hercules graphics standards, PostScript, LaserJet, Epson FX and Toshiba P3 compatible printers.

For copies, visit your local software dealer, or call Wolfram Research at 1-800-441-MATH.

Other Systems

Apollo DN 3000 and 4000 systems: from $2,400
DEC VAX VMS and ULTRIX, and DECstation: from $2,400
Hewlett-Packard 9000/300 and 800 systems: from $2,400
IBM AIX/RT systems: from $2,400
MIPS systems: from $2,800
NeXT bundled as standard system software
Silicon Graphics IRIS systems: from $2,800
Sony NEWS systems: from $2,400
Sun 3, 4, and 386i systems: from $2,250

Supercomputer and other versions are also available.

Prices listed are for lowest-end members of hardware families only. For complete pricing information, and to order copies, call Wolfram Research at 1-800-441-MATH.

Educational discounts are available. All prices and specifications are subject to change without notice. Prices listed apply to U.S. and Canada only.

Software Packages from This Book

The packages described in *Programming in Mathematica* are available for $15.00 on Macintosh and MS-DOS diskettes from Wolfram Research.

Call with your VISA/MasterCard number, or send a check to:

Wolfram Research, Inc.
P.O. Box 6059
Champaign, IL 61826-6059
217-398-0700
fax 217-398-0747

Mathematica.